Fire as an Instrument: The Archaeology of Pyrotechnologies

Edited by

Dragos Gheorghiu

BAR International Series 1619
2007

Published in 2016 by
BAR Publishing, Oxford

BAR International Series 1619

Fire as an Instrument: The Archaeology of Pyrotechnologies

COVER IMAGE
*Replica of a brazier from Gumelnita tradition, Chalcolithic, Cascioarele,
Romania, Made by Andreea Oprita, Vadastra campaign 2004.*

ISBN 9781407300313 paperback
ISBN 9781407330860 e-format
DOI https://doi.org/10.30861/9781407300313
A catalogue record for this book is available from the British Library

BAR Publishing is the trading name of British Archaeological Reports (Oxford) Ltd.
British Archaeological Reports was first incorporated in 1974 to publish the BAR
Series, International and British. In 1992 Hadrian Books Ltd became part of the BAR
group. This volume was originally published by Archaeopress in conjunction with
British Archaeological Reports (Oxford) Ltd / Hadrian Books Ltd, the Series
principal publisher, in 2007. This present volume is published by BAR Publishing,
2016.

BAR

PUBLISHING

BAR titles are available from:
      BAR Publishing
      122 Banbury Rd, Oxford, OX2 7BP, UK
EMAIL info@barpublishing.com
PHONE +44 (0)1865 310431
FAX +44 (0)1865 316916
      www.barpublishing.com

Contents

Acknowledgements

The author thanks all the contributors for the long patience during the preparation of the present book.

Last, but not least, thanks to Cornelia Catuna and Bogdan Capruciu for the invaluable help in editing and correcting the translation of the texts.

D. G.

Contributors

Dr. Ulla Odgaard
Sila - The Greenland Research Centre at the National Museum of Denmark
odgaard@archaeology.dk

Silje Evjenth Bentsen
Cultural Heritage Management Office of Oslo, Pb 2094 Grünerlokka, 0505 Oslo, Norway
siljeeb@yahoo.no

Dr. Judit Regenye
Laczkó Dezső Museum, Veszprém, Hungary
regenyej@mmuzeum.hu

Dr. Jacqui Wood
Saveock Water Archaeology, Saveock Mill, Greenbottom, TR4 8QQ, Cornwall, U.K.
Jacqui.wood@gxn.co.uk

Professor Dragos Gheorghiu
Centre of Research, National University of Arts Bucharest, 19 Budisteanu, Bucharest, Romania
gheorghiu_dragos@yahoo.com

Dr. Claude Sestier
2 promenade F.Rabelais, 77186 Noisiel, France
sestier.flint@wanadoo.fr

Dr. Jes Martens
University of Oslo, University Museum of Cultural Heritage, POBox 6762 St Olavs plass, NO-0130 Oslo, Norway
jesmart@online.no

Dr. Sariel Shalev
Haifa University, Mount Carmel, 31905 Haifa, Israel, & Centre for Archaeological Sciences, E.S.E.R. Weizmann Institute of Science, 76100 Rehovot, Israel
shalev@wiesemail.weismann.ac.il

Dr. Stanislav A. Grigoriev
Institute of History and Archaeology, The Ural branch of the Russian Academy of Sciences, Chelyabinsk, Russia
stgrig@mail.ru

Seth A. Schneider
Department of Anthropology, University of Wisconsin-Milwaukee, Milwaukee, Department of Archaeology, Sabin Hall 290, PO Box 413, Milwaukee, W 53201, USA
sethas@uwm.edu

Bartłomiej Szymon Szmoniewski
Institute of Archaeology and Ethnology, Polish Academy of Sciences, Cracow Branch, Ulica Slawkowska 17, 31-016 Krakow, Poland
bartheque@yahoo.com

Andrzej Kielski
Academy of Material Science and Ceramics, AGH University of Science and Technology, Al. Mickiewicza 30,
akielski@uci.agh.edu.pl

Maria Lityńska-Zając
Institute of Archaeology and Ethnology, Polish Academy of Sciences, Cracow Branch, Ulica Slawkowska 17, 31-016 Krakow, Poland
maria@archaeo.pan.krakow.pl

Krystyna Wodnicka
Academy of Material Science and Ceramics, AGH University of Science and Technology, Al. Mickiewicza 30

Professor Ralph Rowlett
University of Missouri-Columbia, Columbia, Missouri 65211, USA,
ralph.rowlett@gmail.com

Dragana Mladenovic, MSt
University of Oxford, St Hugh´s College, OX2 6LE, Oxford UK
dragana.mladenovic@archaeology.oxford.ac.uk

Marta Caroscio
martacaroscio@interfree.it

Steady old Väinämöinen
stepped out himself, went
up over the wilds
after that harsh fire
and he found the fire
beneath two stump roots
in an alder log
tucked under a rotten stump;
and there old Väinämöinen
put this into words:
"Dear fire, God's creature
the Creator's creature, flame
you had no cause to go deep
and no business quite so far!
You'll do better when you go
back to the stone hearth
commit yourself to your dust
peter out to your embers
to be kept by day
in the cook-house birch-faggots
and hidden by night
within the glowing fireplace."
And he snatched the spark of fire
Put it on burning tinder
on hard birch fungus
in a copper pan
carried the fire in the pan
on the birch bark conveyed it
to the misty headland's tip
to the foggy island's end
and the cabins got their fire
back, the rooms their flame.

(Lönnrot 1999: 632-633)

Lönnrot, E., 1999, *The Kalevala*, translated by Keith Bosley, Oxford: Oxford University Press.

Introduction

Dragos Gheorghiu

The area of the study of pyrotechnologies covers a vast domain that comprises various archaeological topics, including landscape studies, ceramic studies, archaeometallurgy, glassmaking studies, kilns and other pyroinstruments, systems of heating, systems of lighting, deforestation, food preparation, food conservation, funerary and daily rituals, as well as techniques of war.

As a consequence of its instrumentality, in all these instances, fire can be perceived not only as a phenomenon but also as a material artefact, especially when it was associated with a material support, forming together a pyroinstrument. From a semiotic perspective one can say fire has an indexical presence, being represented by the sub-products of combustion, the material supports (like kilns, moulds, lamps, etc.), the temperatures measured in materials, but also by the economic problems generated by its controlled utilization.

The purpose of the present book, which gathers together some of the papers presented at the 9th EAA Meeting in St. Petersburg in 2003, with new contributions, is to offer a "material" perception of fire which will be approached as an artefact, together with its material support.

The history of humankind reveals a gradual emergence of the use of fire as a perfectible instrument, a portable instrument allowing territorial expansion and the colonization of the cold regions like the northern parts of Europe or of America, or the torrid lands of Australia.

Prehistory was the first age of fire, when it was utilized as an instrument for modelling the landscape, the fields or the settlements, for processing materials and for religious use. During prehistory, as in the epoch which followed it, fire was employed to transmute clay into ceramics, to prepare plaster and resins, to smoke food, as well as in a myriad of other tasks. It seems that the oldest evidence for the instrumentality of fire comes from Koobi Foora in Africa, a site dated 1.6 myr, where fractured stone tools are indexes of the use of fire's effect to break hard stones.

The Palaeolithic Era can be regarded as having been an evolutionary step in what concerns the mastery of fire; the first pyroinstruments using encapsulated fire in basin-like clay containers could be identified as early as the middle Aurignacian sequence, i.e. 34-32 kyr BP at Klisoura cave 1, in north-western Peloponnesus, Greece.

Similar to the conquest of the cold, fire allowed the conquest of the night and darkness. As a lighting instrument, fire extended the time of activity of humans, by generating a new, pyrocentred, anthropology. It was also used as an instrument for animation, of transferring *anima* to cave images; the most explicit animated images come from the Chauvet cave in France, where a bull with eight feet, a deer with five heads or rhinos and felines drawn in rows could have been visualized in movement by moving the source of light in front of them.

The Neolithic led to the encapsulation of fire in different ceramic containers (from the small braziers and ovens to the large wattle and daub rooms), the prehistoric *oikos* being in fact an assembly of pyroobjects with diverse functions, from alimentary to religious.

It was then that a better control of fire, by controlling the air-flow, developed; the result was higher temperatures and, consequently, a better quality of ceramic, and later the production of stoneware and porcelain. A new material, metal, followed in time, creating a new specialization which generated fundamental economic and social transformations. The Chalcolithic represented the moment of balance between two main pyrotechnologies, ceramics and metallurgy, with enormous social and technological effects in the epochs that followed. Metals, which replaced ceramics as social prestige objects in the Bronze Age, needed an improved pyrotechnology and conferred on the metallurgist a special magic status, due to the transmutation of materials. Innovative material supports for fire, like furnaces with natural or artificial air-draughts, or devices like bellows, allowed the reaching of very high temperatures for melting copper and iron due to the improved air-draught. **(Fig. 1)**

In the Indo-Europen cultural area metallurgy had its own gods. The *Rig-Veda* (Eliade 1977: 282, n.1) shows the importance of the fire-god for archaic societies, while the study of the positioning of the fire within the separation phase of the rites of passage demonstrates the transforming and exclusion role of fire on materials or persons. Sacrifice through fire, an operation with sacred value, was to be applied at the same time to the human body, architecture and objects, to cite only the custom of firing houses or settlements prior to abandonment, specific for south eastern Europe in the Neolithic and Bronze Age, a practice I believe to have been also a method of purification and of consolidation of the foundation ground for the following overlapping levels of construction. **(Figs. 2 - 3)**

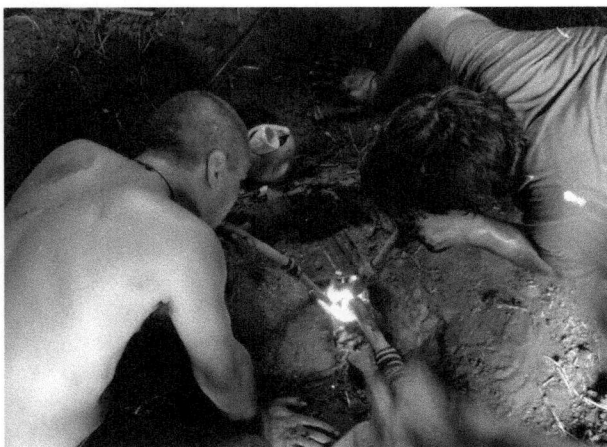

Fig. 1: Experimental copper smelting with air-draught produced by bowels (centre) and two blowers, Vadastra campaign 2005. Performers from left to right: Marius Stroe, Catalin Oancea and Dragos Manea. Photograph by Adrian Stoicovici.

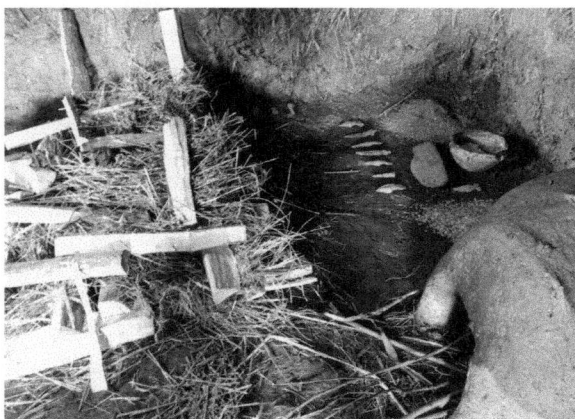

Fig. 2: Two pyroinstruments in a reconstructed Chalcolithic house: left a pyre, right a clay oven. Photograph by Gelu Serseniuc. [Director of the project Professor Dragos Gheorghiu, builders of the house: Catalin Oancea, Marius Stroe, Dragos Manea, and Stefan Ungureanu, Vadastra campaign 2005. The pyre is Dr. Romeo Dumitrescu's experiment of corpse combustion in a fired house; the objects positioned in the background are part of Dr. Fabio Cavulli's (Trento University) experiment of study the effect of fire on the artefacts left in a fired house, Vadastra campaign 2006.]

As an instrument to transform the human body, fire transforms itself into an instrument of passage, an instrument of transport to the Otherworld, being at the same time an instrument for contacting the ancestors. It was at the same time an instrument to guarantee the memory of the place and the conservation of the shape of perishable materials.

The present book studies some of the uses of fire as an instrument from prehistory to the Middle Ages, offering a variety of subjects of investigation, as follows: methodological approaches to prehistoric hearth features, the hearth as a sum of stages within the *chaîne-opératorie* or artefacts, similarities between simple and advanced pyroinstruments, fire as an instrument of memory and a conservation of the ephemeral, the household as a complex pyroobject, the result of the sum of different small pyroinstruments, the function of prehistoric Bunsen lamps with perforated walls, the emergence of metallurgy, iron as the result of the improvement of copper smelting air-draught furnaces, coin minting, traces of fire as indexes for domestic and cult use, pyrorites of passage, roasters for drying grains, and methodological approaches to medieval hearth features.

Fig. 3: The ceramic oven, conserved by fire after the house's combustion. Vadastra campaign 2006. Photograph by Dragos Gheorghiu.

CHAPTERS' DESCRIPTION

The most important pyroinstrument of any prehistoric household for cooking and room heating was the hearth. These pyroinstruments create diverse sorts of heating and their function can be interpreted, thus adding information on the space heating and culinary processes that took place in prehistoric dwellings. **Ulla Odgaard** offers a methodological approach to study hearth features in general and presents theoretical reconstruction of the function of hearths within three Stone Age dwellings, suggesting that hearths should be regarded as structures composed by different elements, which can be described separately, making it possible to interpret their interrelated function. By deconstructing the pyroinstrument into basic elements, one can analyse and compare them with examples from the archaeological or ethnographic record. Biomass combustion technology, ethnographic hearths and archaeological experiments form the basis for the interpretation.

For **Silje Evjenth Bentsen**, a pyroinstrument, like any other object, is the result of processes of construction-deconstruction and abandonment. A hearth is not only a sum of operations and activities, but also an assemblage of artefacts. It is an instrument which necessitates special types of stones for building, a daily activity of fuelling, with wood, bone or domestic waste, and a daily activity of maintaining and relighting. The analysis of the types of fuel used could produce data on the "community's choice", i.e. the cultural preferences of the community. Some activities like the reuse of the hearth create problems of interpretation, because of its specific operations like periodical cleaning or changes of the soil's characteristics. Because it represents the main symbol of the household, the hearth could function also as a social attraction for people as well as for artefacts and its location within the spatial organization of dwelling could provide data on its main function, on the social practices or on the status of its owner.

Ethnographic observations contradict the classical assumption that there is a difference between firing into a bone firing and a kiln; they reveal the mastery of the persons who manipulate fire, a fact which is supported by the paradoxical results of equivalent effects in "primitive" instruments of combustion and in "complex" ones. According to **Claude Sestier** the shape of the pyroinstrument is of major importance for modelling fire and heat, but in the end the skill and experience of the performer are the determinant. When looking for a relationship between the quality of pottery firing and some structures of combustion, one can conclude that it is not the material support which influences the firing process, but the very management of fire, the manner of how it is controlled. Experiments also support the idea that the instrumentality of fire is not completely dependent on the material support (as bone firing or pit-kiln), but on the mastery of the pyroinstrument.

The transformative character of fire, employed by Neolithic populations as a preserving agent, is discussed by **Judit Regenye,** using the architectural remains from the Lengyel culture, Transdanubia, dated to the Late Neolithic and Early Copper Age. Among the materials used by prehistoric communities, clay was the most suitable for preservation, after being transformed by fire, saving most of the cultural information. Regenye discusses the relationship between fire and garbage, i.e. the discarded ceramic fragments, which forms a sort of indestructible information which had a sort of dynamism within the archaeological record, compared with the fragments of daub resulting from the demolition of the houses that are discovered in their original location. Wattle and daub architecture is a good example of how fire can conserve information about building methods and the spatial position of architectural elements by preserving the imprints of the wooden elements within the fired daub. Fire could be perceived as an instrument of memory for fixing in time the ephemeral.

South-east European Chalcolithic material culture is characterized by the analogy of the process of air-draught generated in different ceramic pyroobjects with perforated surfaces (as fire starters and preservers of ambers, ovens with vents and ceramic shutters, liquid heaters, seed or salty solutions desiccators, food smoking containers, braziers in the shape of perforated cylinders or rectangular prisms, sometimes designed to symbolize architectural objects) or in interrelated assemblages (as the macro-pyroobjects formed by the functioning of all the pyroobjects of houses. One of the macro-pyroobjects is the house itself during the process of intentional firing of the architectural structure. According to **Dragos Gheorghiu** such specialized design is the result of the emergence of a new kind of *oikos* as a complex pyroobject, the result of an ensemble made of objects-functions-symbols that were in a holistic relationship of function and meaning. This ensemble defined the living space as a place of sieving and modelling fire and flour, which could also be transformed into an instrument of combustion. The paper insists on the determinism of the forms-functions for the instrumentality of fire.

In the archaeological record of many Neolithic and Bronze Age traditions there is evidence of cone-shaped ceramic objects with perforated walls whose function was identified as being cheese moulds, although their shape would not allow an efficient use for this purpose. It is the merit of experimental archaeology to have offered new interpretations for the use of these peculiar objects as pyroinstruments. Experiments have revealed that the perforated surfaces had the role of suction of air, and that these objects could more likely have been used in a relation with fire than with food substances. By introducing fuel like fat or wax-soaked rush within the aperture positioned on top of the cone-object, a controllable flame could be produced, and an improved air-draught can be generated by allowing air to penetrate at the base of the object. These small ceramic objects function in a manner comparable to the large up-draught kilns with platform, due to the similar way of aspiration and the production of a flame at the upper part of the perforated surface. **Jacqui Wood** experimented with several of these pyroinstruments with perforated walls which she calls "Bunsen lamps", comparing the effects of different shapes and suggesting different functions for each of them, like instruments for soldering weak metals or smokers.

One of the important effects of the mastering of fire in prehistory was metal casting. Slag and crucible remains, sometimes associated with ore, are indexes for a new instrumentality of fire, since the manufacture of these objects implies an association with fire. The emergence of metallurgy represented not only a new specialization but also complex economic problems of trade in ores before being were transformed into metal objects. **Sariel Shalev** discusses the production of the pieces of copper from 5-4 BC millennia found in Ein Assawir in the Near East, which infers that the technology of transforming the cuprite or malachite ore used open clay or sand moulds as well as cold working. It is significant to perceive the crucibles as important pyroinstruments in the process of the mastery of fire. The Chalcolithic process of the manufacture of copper using hot and cold cycles implied a good knowledge of the properties of the metals, by which means selective forging could have produced items of unalloyed metal with a high resistance. In metallurgy fire was a complementary instrument to the process of the formation of metallic items, being cyclically used in the process of the formation of the objects.

The beginning of iron metallurgy is one of the principal topics in archaeometallurgical studies and the examination of pyroinstruments and of the ores used can produce new information on the subject. **Stanislav Grigoriev** stresses that the iron production of the Late Bronze Age Northern Eurasia was possible when metallurgists smelted chalcopyrite, and that a large part of the earliest iron objects produced in the Near East are the result of the metallurgical process of smelting copper ore and do have not a meteoritic origin. The metallurgic technological processes that can be identified at the beginning of the Late Bronze Age in the Volga-Ural region demonstrate the relationship of this region with the Near East and with Central Europe, which puts in a new light the migration of populations with the technology of iron production, to cite the Sintashta tradition. A new type of furnace, present in this tradition and in later ones, formed by small cupola-shape attached to a well, allowed additional air blowing and the raising of the temperature inside the pyroinstrument, permitting the extraction of the metal from silicate rocks and sandstones. The Volga-Ural region had a significant role in the diffusion of the innovative technological know-how.

A particular example of archaeometallurgy is coin minting. **Ralph Rowlett** and **Dragana Mladenovic** examine the coinage made with the help of various furnaces and smelters for the chiefs of the Treveri tribe from the Iron Age oppidum Titelberg in the southwestern Luxembourg, La Tène tradition. These pyroinstruments were used for a very long time and continued for almost a century after the Roman conquest, offering information also on Roman coin minting practices too. Money was formed beside small metallic cult objects, which indicates a wider range of artefacts created in the oppidum, and at the same time a specialization, since the principal production of metal was done in a different part of the settlement. The *chaînes-opératoires* of pyrotechnology involved several stages, to cite only the heating of clay moulds (made of crushed lava mixed with clay) and the repeated rise of the temperature to 1300° C.

Research in favour of an interpretation of some Norwegian sites as being places of cult connected to the use of fire, dated to the Bronze Age (Montelius II-V), was carried out by **Jes Martens** at Glumslöv. The indexical character of fire was revealed in the case of cooking pit sites by the phosphate left in the soil, which can provide data on the difference in use in different areas, and infers a specialization in the use of these pyroinstruments. **Martens** proposes a dual hypothesis for interpreting these structures of combustion, firstly as simple pyroinstruments for cooking, and secondly as instruments of cult, which consumed the offal from animal or human sacrifices. In the second instance the spatial organization of the cooking pit site at Glumslöv would have been perceived as a sacred place bordered by a boundary marker and aligned cooking pits whose principal function would have been to mark the sacred inner space made visible in this way also by the smoke and steam produced. In such case fire was an instrument with a double role, practical and ceremonial at the same time, acting as a ritual item.

Uses of fire as an instrument used in mortuary contexts for defining social differentiation and ritual activities that connected the living population with the ancestors could be observed in the early Iron Age burial mounds in the Speckhau group associated with the Heuneburg hillfort on the Danube River in southwest Germany. **Seth Schneider** discusses the instrumental employment of fire for high status individuals in Early Iron Age mortuary contexts by putting into evidence the relationship between funeral pyres, cremation burials and further pyroactivities in burial mounds as a way for the elite to maintaining their social, economic and religious power. Based on a corpus of ethnographic and historical evidence, the paper extends the cultural role of fire from the functional to the spiritual level, with the attributes of purification, of transporting material to the Otherworld (a pyrorite of passage) and of connecting the living population with the ancestors. The pyrotechnic ritual activity that occurred in the funerary mounds has also a rhetoric discourse, especially for the Late Hallstatt people who developed a form of ancestor veneration using synecdoche in mortuary activities as part of a visual programme of elite signalling. In this context fire was used as a social instrument of legitimizing power.

Sometimes it is difficult to identify the function of a pyroinstrument, an example being the clay roasters which are a special type of vessel with a basin-like shape, found on Early Medieval sites in south, east, and central Europe. Roasters were discovered in relationship with pyroinstruments like hearths, therefore a functional relationship between them and these domestic structures of combustion was inferred. By analysing the temper and firing of the clay vessels, **Bartłomiej Szymon Szmoniewski, Andrzej Kielski, Maria Lityńska-Zając,** and **Krystyna Wodnicka** believe that their major function was the thermal processing of the grains, a practice which would have dried and better conserved grains in pits. Dried grains which resisted longer because their germination was stopped up could have been stored better in pits, an action which had major social and demographic consequences. The analysis of the object's shape infers an undemanding transportability as well as an effective storage. Such a simple form probably allowed, besides the conservation of food with the help of fire, other uses within the medieval household.

A complex instrument like a kiln should be analysed not alone but within its geologic and cultural context, therefore its study will provide data also on the production and manufacturing processes of the epoch. Because of the multifunctional character of such a pyroinstrument, its utilization within medieval society was large, including firing bricks, pottery, and burning limestone. For **Marta Caroscio** a medieval kiln is not only a simple architectural pyrostructure with vaults or pillars depending on the basic function of the kiln, but a compound comprising materials and spaces, like the clay pits, spaces for modelling, or for storing the finished products, which should be perceived in a historical perspective For example, the replacement of stone by bricks in 11th century Tuscany cannot be understood without the study of the kilns within the economic background and the salary system of the time, together with the study of the local natural resources.

Bibliography

ELIADE, M.
   1977 *From Primitives to Zen. A Thematic sourcebook of the history of religions*. San Francisco: Harper and Row.

Hearth, Heat and Meat

Ulla Odgaard

Introduction

To the eye of the modern beholder a prehistoric hearth is often nothing more than a heap of blackened rocks and dirt. But for the archaeologist it is an important starting point for an understanding of the practical functioning of heating and culinary praxis.

Understanding of the pyro-technologies can play active roles in the archaeological reconstruction of life at a prehistoric site or in a prehistoric dwelling. The present paper offers a methodological approach to study hearth features in general and presents interpretation of the function of hearths within three Stone Age dwellings.

Method

Comprehension of the function of an archaeological hearth calls for knowledge about hearths in use. Anybody who has experienced lighting a fire knows that it requires expertise to make it burn quickly and well. Technical analyses of combustion processes and ethnographic examples of hearths in use can provide us with this knowledge.

To deduce the practical function of prehistoric hearths normal archaeological comparative analysis can be applied. Here I will focus particularly on the function of hearths for space heating and culinary practises. The basis is not a descriptive typology. Instead I have chosen to see hearths as composed by elements, which can occur in different combinations. The combinations of the elements can be analysed and compared to other archaeological or ethnographic hearths, which can illustrate aspects of the function.

Combustion processes

Also today people in many parts of the world are using open fires for cooking and room heating. A conference in 1981 had the purpose to investigate the third world's fuel situation, and it was estimated that about half of the world's population lived from food, which had been prepared over an open fire with fuel of wood, agricultural waste or animal dung (Clarke 1985). The consumption of fuel is a major problem and has in many places been the direct cause to destruction of the bio-climate with disastrous consequences. For that reason researchers have made detailed studies of "primitive combustion processes" with the purpose of optimising the efficiency (Stewart *et al.* 1987).

A bio-mass combustion process can be described in four steps, but all steps can take place simultaneously. A piece of wood for example is not burnt in one step. And a fire, supplied with wood, during for example cooking, will go through different phases of combustion simultaneously (Stewart *et al.* 1987).

Step 1: At the ignition – hereby supplying the start-heat – the contents of water in the fuel will evaporate at about 100°C. During this step the fuel absorbs the heat energy.

Step 2: When the heating temperature reaches between about 200° and 350°C the volatile gases (carbon, hydrogen and oxygen) are released.

Step 3: The volatile gases are mixed with free atmospheric oxygen and ignite at temperatures above 450°C and burns with a yellowish flame, radiating heat. Some of the heat is re-absorbed by the fuel and release more volatile gases. This process should keep itself going until all the volatile gasses are released. The volatile gases need sufficient heat, oxygen, space and time to ignite. If any of this is absent the volatile gases can leave the combustion area without being ignited. In this instance the combustion will be unfinished and inefficient, and the fire will give out smoke and die out. A slightly turbulent flow of air will make the volatile gases and oxygen mix better. When all the volatile gases are released charcoal (consisting mainly of reinforced carbon) is left.

Step 4: Charcoal is burnt (oxidizes at temperatures around 800°C), if enough oxygen is present in the fireplace. The produced carbon monoxide reacts with oxygen right above the combustion area (if enough oxygen is present) and emits carbon dioxide. Usually the charcoal will continue to burn long after the volatile gases have been used up. A charcoal fire needs oxygen at the combustion area (primary oxygen) and right above the coal (secondary oxygen). If the secondary oxygen is insufficient, the fire will give off carbon monoxide, which can be dangerous for the user, especially within a tight room.

A charcoal fire only goes through step 4. If there in a wood fire is much charcoal left after the fire has died out, it is an indication of insufficient oxygen or heat at the combustion area (Stewart *et al.* 1987).

The effect of hearths

Heat is energy. Molecular movement stops at absolute zero (-273°C), but above this temperature molecular movement occurs in every substance. As the substance is heated the molecules will move faster. "Heat transfer" is the transference of this energy from one place to another. With a stove: from the fire to the food, and with room heating: from the oven to the room.

There are three main forms of heat transfer:

1. *Conduction* is heat transfer through a solid mass from a warm space to a cold space.

2. *Radiation* is radiant heat. All bodies radiate heat, and the warmer they are, the more they radiate. The yellow flame produced by the volatile gases radiates heat just as the glowing coal in a wood- or charcoal fire. Radiation is the main cause for heat transfer from the fire to the pot.

3. *Convection* is the heat transfer occurring by means of the flow of a fluid (for example air or water). A warm object placed in a stagnant, colder fluid will result in a movement of this fluid because of a reduction in the density of the heated fluid close to the warm object. This fluid will then move upwards and be replaced by colder fluid (Stewart *et al.* 1987).

Terminology and description of hearths

To be able to interpret the function of a hearth it is necessary that all elements are sufficiently described.

At the colloquium "Nature et fonction des foyeres préhistoriques" in Namur in 1987 it was agreed upon that archaeologists should concern themselves with hearths and activities related to it equally much as they are concerned with analysing the technical and economic consequences of flint production (Olive and Taburin 1989). It was further stated that even though the hearth is an evident structure, archaeological descriptions are often inexact and limited to an overall interpretation. This implicates the need for using a descriptive and precise vocabulary without ambiguities in order to establish solid basis for comparative studies (Olive and Taburin 1989). It is, for example, important to be able to distinguish between the overall "hearth/fireplace" and the more precise "combustion area", since even though a feature is described in the archaeological literature as "hearth" or "fireplace" it is not always obvious where the actual combustion process has taken place.

To comply with these demands and inspired primarily by the work of Perlès (1977) I will suggest that hearths should be regarded as structures composed by different elements, which can be described separately making it possible to interpret their interrelated function.

The elements a hearth can be composed of are:

- Traces of combustion: ash, charcoal, burned bone or fat
- Moveable rocks, possibly fire-cracked
- Feature with area of combustion, which can be a fixed stone construction

Traces of combustion

A combustion process that includes firewood will always leave traces of charcoal and/or ash. The probability/degree of washing out or other factors, which may have removed charcoal from a feature, should be considered already during the excavation situation.

Charcoal should be quantified. The measure can be given by volume, making it possible to estimate the amount in relation to other hearths, as for example done by Soffer (1985) to estimate the intensity in use of different hearths.

Fire-cracked rocks

Within archaeology there is a large yet still relatively unexploited potential in analysis of stone material from hearths with moveable rocks.

The documentation of fire-cracked rocks can be carried out on different levels. Firstly, a simple recording of whether fire-cracked rocks are present or not, their location at the site and in connection to which features. Secondly, weighing of the fire-cracked rocks can be done in order to provide quantification, which can be compared to other features and sites (Olsen 1998). This kind of documentation will, however, not be precise enough to make it possible to estimate the length or intensity of the use of a site, since different types of rocks have different capacities and will be worn down at different rates when used as heating elements. This has been clearly illustrated by Buckley (1990), who through experiments showed (table 1) that sand- and limestone (sedimentary) produces more debris after the same number of heatings and dowsings than basalt and gabbro (igneous rocks). Arkose and agglomerate is only half as quickly worn down (to less than 5 cm) as the sedimentary rocks, but twice as quick as the igneous.

If they are granites it is more complicated. Granite is an igneous rock, which in contrast to other igneous rock types acts heterogeneously when heat-treated. Hard and dense granites are just as durable as other volcanic rocks, while other granites will break after one single heating. For this reason, when the archaeological material consists

Table 1: The two first columns are from Buckley's experiments (1990) showing how many heatings/dowsings it take for a rock to disintegrate into pieces smaller than 5 cm, considered as unsuitable for more heatings. I added the third column.

Rock type	No. of heatings / dowsings	Rock
Micaceous Sandstone	5	Sedimentary
Limestone	6	Sedimentary
Agglomerate	10	Metamorphic
Arkose	12	Metamorphic
Basaltic	20	Igneous
Vesiculated Basalt	>25*	Igneous
Gabbro	>25*	Igneous

* Experiments discontinued owing to no visible upper limit being foreseeable

of granites, it is important to examine the quality of the local granites. One method of doing so is to carry out experimental heatings and dowsings of rocks collected in the area around the archaeological feature (Markström 1996). Markström's experiments showed that granites behave differently according to whether they were fine-grained or coarse- grained, and depending on their quartz content. Breakage did not seem related to whether they were cooled in water or in air. Coarse-grained granites with a high content of quartz broke during the initial heating, and after cooling were filled with micro cracks making them useless for re-heating. Fine-grained granites were almost uninfluenced after heating and cooling, typically exhibiting only a few thin cracks. On the other hand, examinations of the rock's ability to accumulate heat have proved that fine-grained granites loose their heat considerably quicker than coarse granite. It is apparent that rocks that develop cracks after repeated reheating lose the ability to retain heat. This is especially true of friable granites (Markström 1996).

Examination of rocks was carried out in connection to the interpretation of a Palaeo-Eskimo hearth with more than 550 rocks (mostly granite and gneiss) from the Early Saqqaq culture. It was not possible to obtain rocks for experimental heatings from the vicinity of the hearth since this aspect had not been considered when the excavation was carried out. Instead the rocks from the hearth were examined for data (size, rock type), which could be compared with the results of the experiments made by Markström (1996) and Buckley (1990). This analysis made it among other things possible to distinct

three episodes of use during one of which large quantities of meat or fish was dried (Odgaard 2001, 2003).

To be able to tell whether the rocks collected for heat treatment were selected meticulous or picked randomly it is necessary to investigate the frequency of rock types found in the vicinity. At Head-Smashed-In Buffalo Jump in Alberta, Canada, it has been demonstrated that effort and time was spent on importing rocks, since the great majority of the fire-cracked rocks were not local (Brink and Dawe 2003).

Typology

Research into hearths has pointed out some basic problems among others that it is only possible to compare hearths when a preliminary classification exists, including a hierarchic organisation of the descriptive traits of the features. But regarding hearths the descriptive approach may not cover the full sequence of uses and functions a hearth had in the dwelling. What we find at the sites are hearths in a final state, where acts and activities have modified the original state.

If the function of hearths remains unidentified, a classification based solely on the morphological traits can only establish formal distinctions. Consequently, a typology showing different types of hearths and not different states of use can only be established based on an understanding of the use and function of the hearths (Coudret *et al* 1989).

Every hearth should in principle be looked upon and interpreted within its own context. I will however as a starting point suggest a simple typology, with implications regarding interpretation of the use, based on my work with hearths in general (table 2).

Table 2: Hearth typology

HEARTHS:	WITHOUT ROCKS	WITH ROCKS	
		Fixed rocks	Movable rocks
PROCESS:	Open combustion Radiation	Open combustion Radiation Convection	Closed combustion Convection
RESULT:	Light and heat	Light and heat	Heat
CULINARY OPTIONS:	Broiling/grilling Boiling/cooking in pot	Broiling/grilling Roasting Boiling/cooking in pot	Roasting Boling/cooking in pot Boiling with rocks

Hearths without rocks

A hearth – where rocks are neither part of the construction nor form part of it as moveable fire-cracked elements – can, since it does not contain heating elements, only transfer radiant heat to a dwelling, and the combustion process requires good ventilation. The process will yield light, but culinary options are limited to broiling or grilling while direct boiling would be possible only if a fireproof container is available.

Hearths with fixed rocks

A combustion area situated on/within a construction of stone – beyond the radiant heat produced by the open fire – can afford convectional heat from the stones, when the combustion process has come to an end and it is possible to shut off the ventilation of the dwelling by closing the smoke-hole and/or the entrance. The first part of the process will provide for light while the second part will leave the dwelling in darkness unless other sources of light are in use. In addition to broiling, grilling and boiling in a pot, in hearths with fixed rocks it is also possible to roast by placing the food directly on the hot rocks.

Hearths with moveable rocks

Hearths with a content of moveable rocks should be interpreted after considering the relation of the rocks to charcoal.

Rocks that are not in context with charcoal could have been transported from the heating source into a dwelling where all air-channels are tightly closed. The process of heat transmission would be convection, affording an even temperature in the room. Fire-cracked rocks that are clean and found outside the context of sooth and charcoal could moreover have been used for boiling of liquid in a not necessarily fireproof container.

Rocks that are mixed with charcoal were probably heated on the spot in an open fire. When the fire died out the rocks continued to function as heating elements, affording moderate heat for a longer time than the embers. During the process it would have been possible to boil/heat liquid in a fireproof container put directly on or among the rocks, and flat rocks could have been used as frying pans.

Experimentation

Also experimentation plays an important part in the ongoing research on hearths, as does studies of ethnographic documented hearths. Through experimental archaeology questions about specific hearth types can be illustrated (for example Odgaard 2005; Soffer *et al.* 1993). Experiments can control the efficiency of different hearth-arrangements and can throw light on both given conditions and deliberate choices. On this background experimental models can be developed, which in connection with the archaeological context make it possible to distinguish between conditions and choices.

A promising method for the study of hearths is micro morphology. This method implies microscope studies of elements from prehistoric hearths compared to elements from experimental hearths. For example questions of reached temperatures can be investigated by this method (Soffer *et al.* 1993).

Further analyses

More sophisticated technical analyses, as shown by primarily the French archaeologists (Olive and Taburin 1989; Wattez 1988), should be applied to the study of the hearth. It is an obvious area for archaeometric methods, and technical analysis of the construction and content of the hearths can form part in the analysis as information, which can support the interpretation of the practical aspects. But without archaeological interpretation, which includes analogy, research provides just more accurate descriptions of hearths.

We need to synthesize, and I suggest that the methods and considerations presented in the present work can be made use of in this connection.

Deducing the function of hearths

To illustrate the potential I will present interpretations of three Stone Age dwellings with hearths. Two dwellings are from the Palaeo-Eskimo Cultures of Canada and Greenland and one is from the Ertebølle Culture of Denmark. They are chosen because they have been published with good descriptions of the well-preserved hearths and dwelling remains. My aim here is not to give the final interpretation of these. Rather I wish to demonstrate a method for interpreting the functions of hearths as inspiration for future analysis and excavations. I want to emphasize that focus on the hearth structures already during the excavation is important, because at this stage qualified judgement about which analyses to imply is necessary.

Arctic hearths are especially interesting because of the circumstances under which they functioned. Firewood was often scarce, and outdoor temperatures were very often bitter cold and at the same time choices of material were limited. The Palaeo-Eskimo answer to these challenges was a multifaceted pyro-technology (Odgaard 2003; 2001).

Jensen (1998) pointed to a correlation between the decrease in the amounts of fire-cracked rocks at the sites and the introduction of soapstone vessels, which occurred around 1300 BC (Møbjerg and Gotfredsen 2004). This incidence revolutionized the pyro-regime of the Palaeo-Eskimos (Odgaard 2003; 2001). In the following two examples, one from prior to and one from after this shift, will be presented.

1. A High Arctic hearth with small moveable rocks.

The first example is a hearth from the earliest pioneers of the High Arctic, the Palaeo-Eskimo Independence I culture (2500-1900 BC) from Lakeview Site on Ellesmere Island in the Canadian High Arctic (Schledermann 1990).

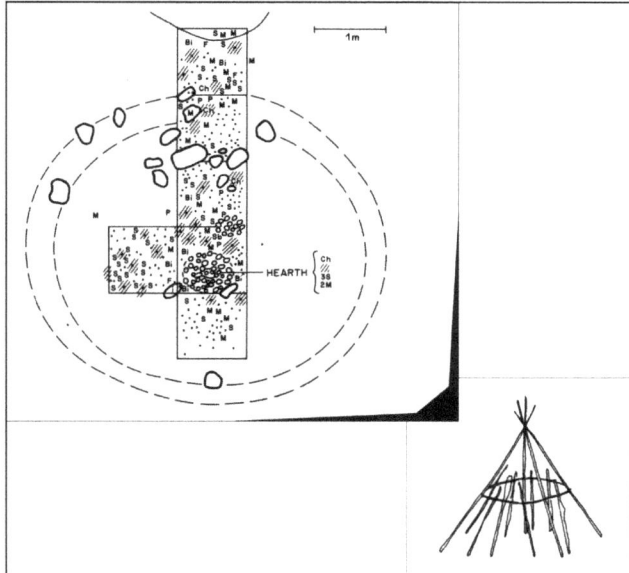

Fig. 1: Palaeo-Eskimo dwelling from Lakeview Site, Ellesmere Island from Schledermann (1990: Figure 10). Illustration of suggested reconstruction from Faegre (1979:132).

The hearth is composed of small, round stones put directly on the exposed bedrock-surface. The combination of the round to oval (4.5 x 3.5 m) tent ring (of gravel) and the hearth type and placement suggests a tipi type of tent, which would make enough room for a central open fire (**Fig. 1**). The fire would provide sufficient light for different kinds of handicrafts, which is reflected by the finding of microblades, burin spalls, flake knifes and points within the tent ring. Now and again an open fire on a flat surface requires cleaning out. In this instance charcoal and ashes were occasionally scattered outside the fireplace itself.

The combustion area of this fireplace was placed directly on the rock – and the combustion process was an open fire heating the tent by radiation. Most of the heat vanished together with the smoke up the smoke hole, suggesting this dwelling to be from a warmer period of the year.

The cooking options were broiling over flames, grilling over embers and boiling. The small round and flat stones in the hearth and in a small pile nearby indicate that cook-stone-technology was applied for boiling as suggested by Schledermann (1990).

Cook-stone technology in culinary practises is known from different ethnographic sources, for example about the Plains Indians who also led a life of limited material

choices. They heated the rocks directly in the fire before putting them into the water. From them we can also learn that all game animals contain a "cooking pot". The stomach is so durable that it is possible to boil in it with heated rocks (**Fig. 2**).

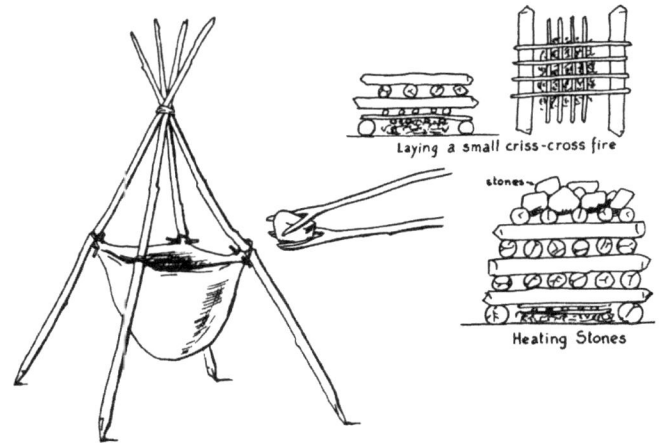

Fig. 2: Boiling with hot rocks in the stomach of a game animal (Laubin and Laubin 1989).

This method of using rocks for boiling fits well with the round hearth from Lakeview Site with the small and round rocks. The rocks being round suggest that they were not fractured by repeated heatings, and that they were probably selected for their present size. Their smallness indicates a heating of liquid as smaller rocks are effective for this exact purpose because of the quick absorption and liberation of heat. Large rocks absorb and liberate heat slower, which make them more suitable for other culinary purposes and for room heating (Markström 1996).

Suggested further analysis for this kind of hearths: A closer study of the small rocks, interpreted as "boiling rocks", could identify the rock type, the size and degree of cracking, which would make us able to estimate the intensity/length of use of the hearth following Buckley (1990).

2. A Palaeo-Eskimo dwelling with two hearths

The second Palaeo-Eskimo dwelling is from Polaris Site, Northwestern Greenland also in a high arctic environment. The dwelling is from the Late Dorset culture (in Greenland ca 750-1300 AD), which is the latest phase of the Palaeo-Eskimo life in the Arctic.

There are two hearths in the dwelling, incorporated in a central "midpassage" (**Fig. 3**), which is a common feature in Palaeo-Eskimo dwellings. The rounded-rectangular shape of the shallow (20 cm deep) housepit, the relatively narrow benches (around 125 cm) at the sides, and the location of the hearths in the broad midpassage make it

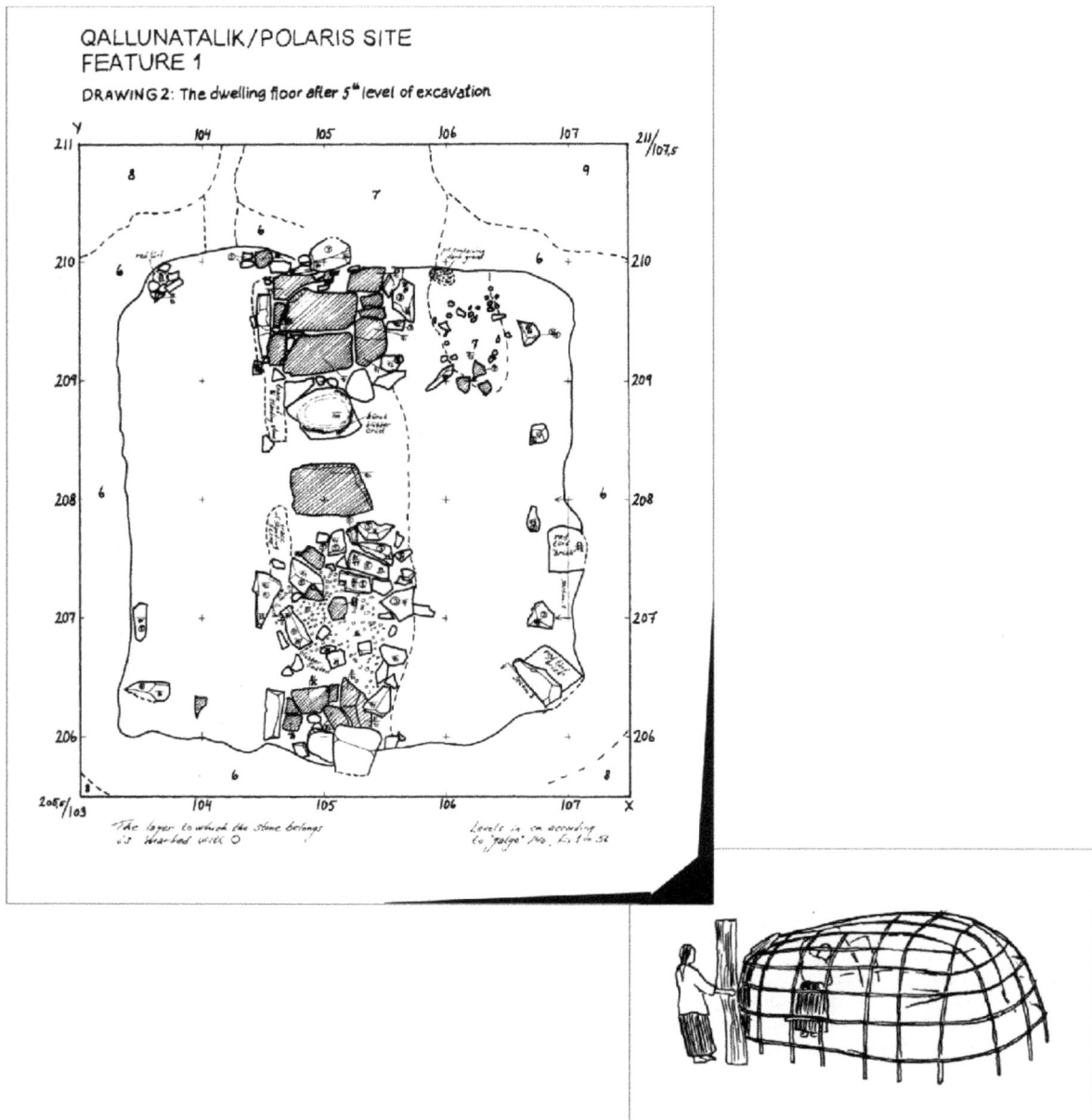

Fig. 3: Late Dorset dwelling in High Arctic Greenland (Grønnow 1999: Figure 44). Illustration of suggested reconstruction from Faegre (1979:140).

likely that the tent was dome-shaped (Odgaard 1995). A probable lamp-support was found at one end of the midpassage. The lamp-support is a heavy rock of ca 30 kg with an even surface and an oval depression rimmed with black-crusted blubber. Also along the sides of the rock the soil was black, probably from overflowingblubber. This could be noted to a depth of 15-20 cm in the soil (Grønnow 1999: 51). The other hearth area is a horizontal 2 cm thick slab 15-20 cm in diameter, with traces of fire-cracking. An 8-10 cm thick layer of pea- to nut-sized round pebbles surrounds the slab. Next

to this arrangement a somewhat cracked stone paving is present. Close to the lamp-support a possible andiron, a rock with at one end the shape of two horns and a sickle-shaped layer of blubber, was found (Grønnow 1999: 52). The Palaeo-Eskimos had by then long since introduced lamps and pots of soapstone, which at this point could reach dimensions up to 50 cm (Maxwell 1985).

The lamp to have stood at the lamp-support represents a prolonged combustion process, which by radiation could provide heat and light, while clothes or other things could be dried above it. If the smoke hole and tent doors were closed this arrangement would provide the tent with a warmer temperature not only close to the lamp but also in the whole tent due to convection. If andirons were placed on each side of the lamp-support they could have supported a cooking-pot – heated by the lamp. The other hearth represents an open combustion facility used for a shorter process in connection with culinary practises with the smoke hole open. Here the moveable andirons could have functioned as support for spits used in broiling over flames or grilling over embers, or without the andirons for roasting on a flat rock with the smoke whole open (**Fig. 4**).

The two hearths in combination provided the dwelling with a source of heat, light, a stove and a multi-kitchen supplying all the needs a Palaeo-Eskimo gourmet could ask for.

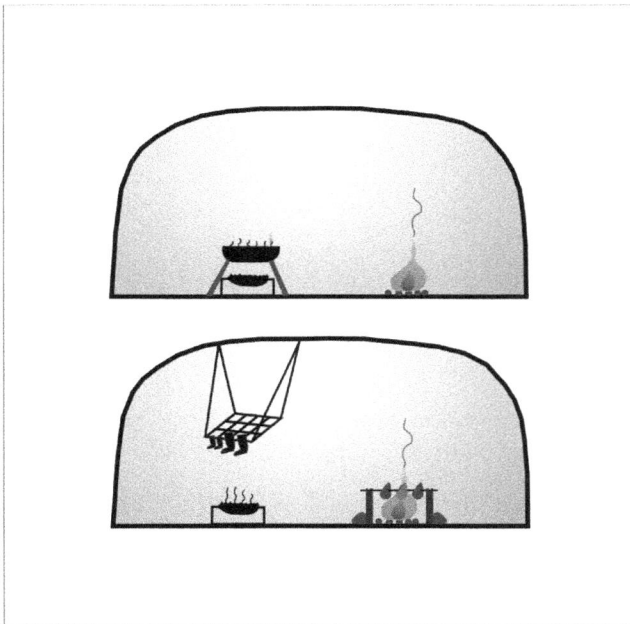

Fig. 4: Reconstruction of hearth arrangements in late Dorset dwelling.

3. An Erteбølle dwelling with three hearths.

At the Mesolithic settlement Lollikhuse on Sealand, Denmark, several hearths were investigated and described with more detail than usual for Danish archaeology (Sørensen 1993a and 1993b). Also a dwelling from the Early Ertebølle Culture with three hearths was found at the site (**Fig. 5**) (Sørensen 1993a and 1993b).

The trace of the dwelling is a shallow pit, 0.2-0.3 m deep, 5.5 m long and 4.0 m wide with several postholes and stake holes along the edge on one long and one of the short sides. The missing stake holes at the other two sides can be due to the applied excavation method (Sørensen 1993a: 20).

In contrast to the Palaeo-Eskimos the environment of the Ertebølle people with a forested landscape presented them with possible choices of dwellings that were not necessarily mobile, since building materials were present at most places. Sirelius (1906) documented a large range of traditional dwelling forms of the Finnish- and Ob-Ugrische area. One plausible analogy for the Lollikhuse dwelling is what Sirelius called an "earth-tent", which people, who lived from hunting and fishing along the rivers Ob and Irtysh in Western Siberia, at the time of Sirelius used for winter dwelling (Sirelius 1906). It was not really a tent since it had neither a mobile frame nor cover. It was built if wood, had a pyramidal form (with pointed or flat roof) with either rounded or square corners at the ground. It was build over a pit, usually between 30 and 60 cm deep, square with 3 – 4.85 m long sides. The height of the dwelling was between 1.67 and 2.15 m (Sirelius 1906: 84f). To keep the dwelling warm there was a thick cover of hay or moos with a layer of earth on top (Sirelius 1906: 85). These dwellings could have an earthen sleeping platform (Sirelius 1906), but also an even floor as the Lollikhuse dwelling.

Fig. 6a shows the principle of the construction of this dwelling-form and a possible explanation for missing post-holes at prehistoric sites: logs of timber as lower base (although Sirelius (1906: 103) considers this to be a late innovation inspired by timber log buildings). The Lollikhuse dwelling has a row of stakes along two of the walls, and another of Sirelius' illustrations (**Fig. 6b**) shows an "earth-tent" with a row of stakes around the base of the dwelling, probably in order to stabilise the construction. The row of close stakes at the western edge of the Lollikhuse dwelling pit could form part of a door-arrangement and a construction that stabilised the rim of the pit at the entrance, where traffic otherwise would trample it. The larger postholes west of the pit can correspond to the either short (**Fig. 6c**) or long (**Fig. 6d**) "entrance halls" described by Sirelius in connection to the earth-tents.

Inside the Lollikhuse dwelling are three hearth areas. Hearth A is in Sørensen's typology: a classical round stone-build hearth of a type known from more or less all classical excavations of kitchen middings in Denmark. The rocks being almost exclusively different types of granites or sandstone are placed as a flat cobbled area (Sørensen 1993b). The particular hearth in this dwelling measures 0.8-0.9 m in diameter. It had been built over an earlier hearth, presumably of the same type, from where some of the rocks had been robbed. Around the hearth were large amounts of charcoal (Sørensen 1993a).

According to the functional typology described above this hearth with fixed rocks gives the options of open combustion providing the dwelling with light from the open fire and heat as radiation from the flames. Most of the heat will vanish together with the smoke, making it an uneconomic room-heating process. The rocks however will accumulate heat, which – after the fire has gone out

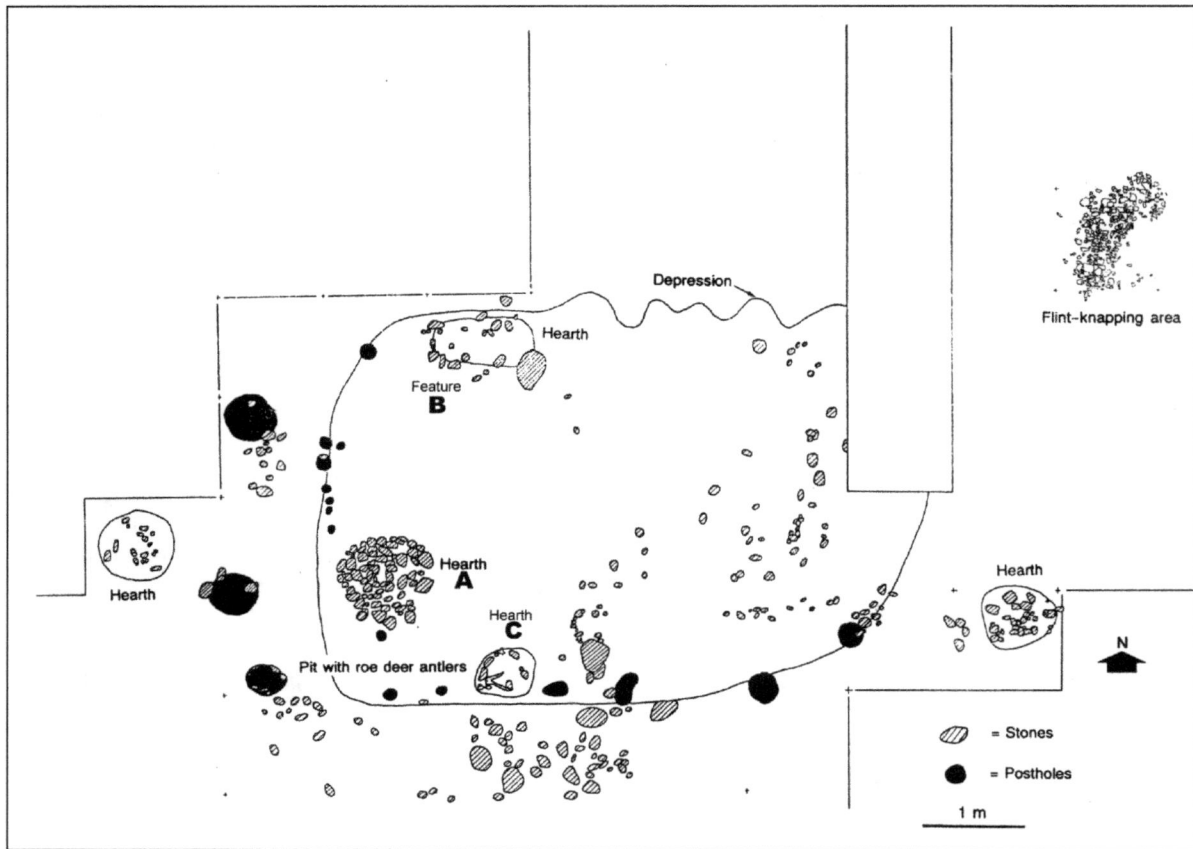

Fig. 5: Ertebølle dwelling with indoor hearths from Sørensen (1993a: Figure 2).

and openings of the dwelling has been closed – can provide with heat by convection. Sandstone would not be suitable for boiling-stones because they are porous. Wet sandstone can explode when heated in a fire, but when dry it has a high thermal conductivity and will warm up quickly.

An ethnographic account describes a dwelling with a hearth that efficiently combines radiation and convection. The hearths of the Mackenzie Eskimos were built of large rocks and had a central position on the floor. Directly above the hearth was an opening in the roof covered by a thin piece of hide parchment and right before the fire was lit the cover was removed. The flames nearly reached the ceiling and created a sudden draft preventing the smoke from spreading in the house, and while the fire was burning a crevice was kept open below the entrance for providing a draft. When the cooking was done the fire was allowed to die out until only a few coals were left. At this time the large rocks around the hearth were warm. Before the parchment was placed to cover the hole in the ceiling again the last coals were carried outside, so no smoke would fill the interior. Even on a very cold day the rocks in the hearth could accumulate so much heat that a kindling of a fire every 6th hour was enough for keeping

a comfortable room temperature (Stefansson 1922: 94 f). The rocks of the hearths of the Mackenzie Eskimos were bigger and therefore accumulated more heat than the hearth in the Lollikhuse dwelling, but the principle of heating by radiation followed by convection could be the same.

The culinary options were broiling over flames, roasting directly on the rocks or on a "frying-pan" or grilling over embers. Because of the dating of the dwelling to Early Ertebølle when ceramic pots were not used, this hearth gives no possibilities of boiling unless in combination with moveable rocks.

A second feature (C) at the northern wall is a rather big patch of grey ash with only little charcoal but with scattered occurrences of fire cracked rocks. It is not clear whether this is a combustion area or it is a secondary placement of burnt material.

If a combustion did take place, this hearth could provide light and heat by radiation and the fire-cracked rocks had a function as heating elements – either for room-heating or for cook-stone boiling in a bark or skin container. Micro-morphology in combination to experimental

6a

6b

6c

6d

6e

Fig. 6 a-e: Examples of the "earth-tent" from Sirelius (1906: Figure 19, 14, 16, 21 and 17).

heating on a surface similar to the ground beneath this feature could reveal if combustion took place and also which temperatures were reached (Soffer *et al.* 1993).

But also the rocks are interesting, because of their possible relation to cook-stone technology and in this instance maybe to boiling. As in the case of the Palaeo-Eskimo hearth with small rounded stones above, it would be an advantage to examine the rocks for their suitability for this purpose.

The aceramic Coast Salish indians according to Batdorf (1990) boiled with hot rocks during the winter when stored dried food was prepared. Round volcanic rocks were heated in the coal of the fire and put in water in a basket or a box of cedar, where the rocks would bring the water to the boiling point. Their fireplaces for heating rocks are described as "small and simple", and it was necessary to use hard wood for creating the intense heat, which only glowing embersproduce. The Coast salish indians placed the rocks close to the fire and on top of

BOX COOKING –

② STONES ARE RINSED OF ASHES BY DIPPING INTO SMALL WOODEN BOX OF WATER.

① BOILING STONES, HEATED IN FIRE, ARE REMOVED WITH SMALL FIRE TONGS

BOX COOKING, IN GENERAL USE THROUGHOUT THE NORTHWEST COAST –[WATERTIGHT BASKET ALSO USED FOR COOKING –].

③ STONES ARE THEN PUT INTO LARGE BOX WITH WATER IN IT.

④ WHEN WATER BOILS, FISH PIECES ARE PUT IN, SOMETIMES CONTAINED IN AN OPENWORK BASKET.

⑤ BOX IS COVERED WITH MATS TO HOLD IN HEAT AND STEAM.

FISH PIECES ROASTING BY FIRE.

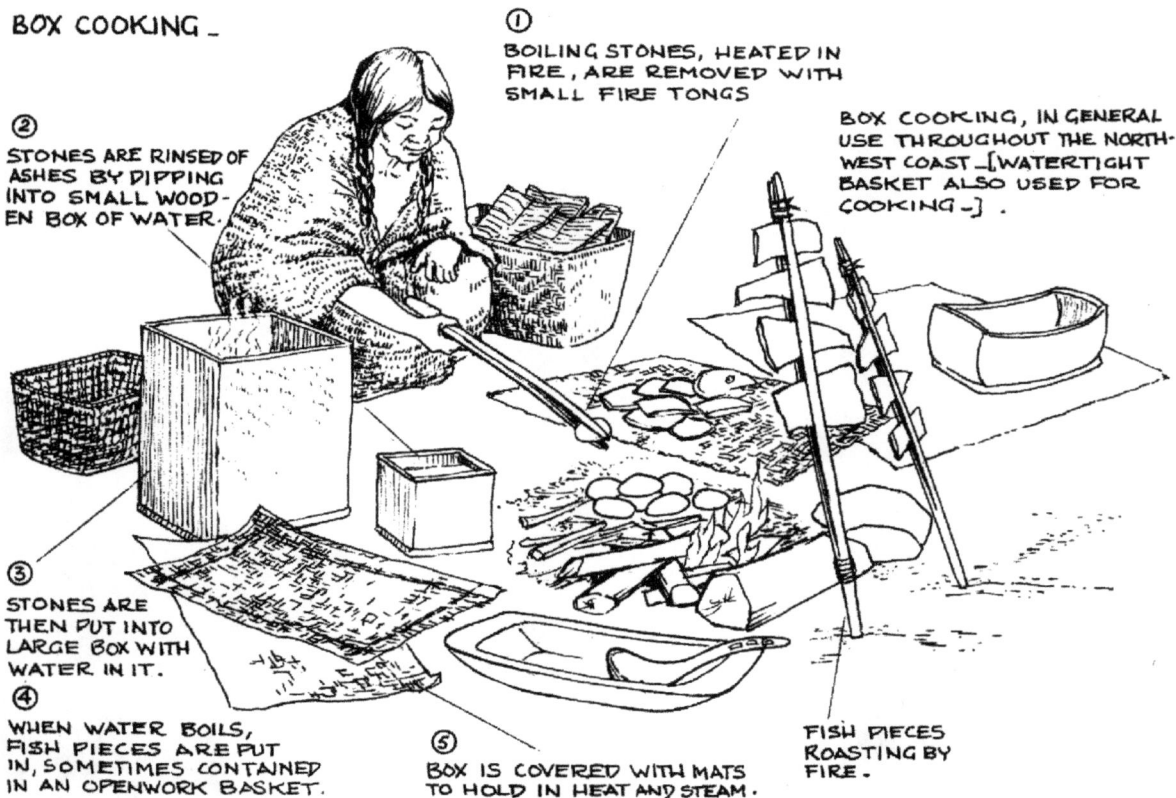

Fig. 7: Without any knowledge of pottery, coast Indian peoples boiled, simmered, steamed, baked, toasted and roasted most of their food (Stewart 1977: 129, figure p. 130). Drawing from Indian Fishing, Copyright © 1977 by Hilary Stewart. Published by Douglas a McIntyre Ltd. Reprinted by permission of the publisher.

other large and flat rocks in order to keep the boiling stones as free from the ashes as possible. Hearth A in the Lollikhuse dwelling with its pavement of rocks would make an excellent surface for a similar process, where the rocks could be heated on the warm pavement of sandstone and granite next to a small intense fire. The Coast Salish Indians appreciated cleanliness. The rocks wereconstantly rotated and moved around for steady heating, thrown into the basket and removed again to be heated in the fire after cooling. Before putting the rocks into the basket they were quickly dipped in another basket with clean water without reducing their heat considerably (Batdorf (1990). Inspiration for a theoretical furnishing of this area of the dwelling can be found in the Coast Salish kitchen area that like the early Ertebølle was aceramic but still well equipped with materials from a forested environment (**Fig. 7**).

In a distance of only ca 1m from hearth A another hearth (B) is located. This structure is a pit ca. 0.1 m deep with a dark brown fill quite rich in charcoal, particularly at the base which in places had a reddish hue due to the heat (Sørensen 1993a: 23). The pit was edged with rocks up to 0.15 m in diameter in a circle 0.4 – 0.5 m in diameter.

Within this structure there was also an intact roe-deer antler and another similar one was found in the central part of the dwelling (Sørensen 1993a: 23).

Hearth B – although in a depression – would give the same options for processes as hearth A: Light and heat by radiation – except that the rocks are too few to accumulate heat of any significance. Rather they seem to have encircled and helped to intensify the combustion process in the shallow pit. Culinary options were broiling over flames and grilling over embers. The heat of the embers was concentrated by the placing in a pit making this structure a perfect grill. The roe deer antlers, either alone or placed on each side of the pit with the prongs pointing upwards, could have served a function as holders for spits as in the case of the possible andiron in the Late Dorset Palaeo-Eskimo dwelling. At upper Palaeolithic sites in Russia, fragments of tubular bone from large animals – especially mammoth – have been found placed diametrically on each side of a hearth with the epiphyses pointing upwards. Sometimes the fragments are of mammoth tusk, and they never bear any trace of burning (Perlès 1977: 81). Same kind of arrangements in stone has also been observed and suggested to be perfect arrangements for roasting on spit (Perlès 1977: 81) and the antlers at Lollikhuse could fit with the same interpretation.

Grilling of meat produces much smoke that could explain the placing of the arrangement close to the wall where another entrance/window could have had its location, as is often the case in the earth-tent (e.g. **Fig. 6b**). At Lollikhuse this is the southern side of the dwelling, which would be an optional placing of an opening.

Another possibility for hearth B is, that the construction around this combustion area was a parallel to Sirelius' "kochofen" or "kamin" (Sirelius 1906: 102). Such a stove was in Western Siberia sometimes located in the middle of the floor but could also have its place along one side of the dwelling as in Lollikhuse. It was build of thin wooden poles to form a cylinder with upwards decreasing width (fig. 6e), and it was tightened with clay mixed with hay, but other materials such as hide could be suggested for the Lollikhuse dwelling. The part of the oven being inside the earth-tent made a fireplace with a large opening and the outside part could be a kind of chimney (fig. 6d) (Sirelius 1906: 85f.) or a square opening in the roof, where smoke could get out and light get in (fig. 6b) (Sirelius 1906: 102). An obvious advantage is the possibility of a more efficient and safe combustion process during control of (secondary) ventilation for air and smoke during longer combustion processes, in this way securing that carbon monoxide is not emitted. The location of hearth B only 1 m from hearth A would make it possible to operate the processes in both structures from a work space east of hearth A and north of hearth B, especially if the location of the fire wood was in the corner between the two structures.

Whether this hearth was a grill or an oven (or both), this must be the place where the fire was kept alive, e.g. during night, in the form of glowing embers buried in charcoal and ashes. When the coals were not protected well enough against draft a new fire had to be made, which was probably done by percussion or friction of stone on stone (e.g. flint on pyrite) (Stapert and Johansen 1999). Both in Greenland and Europe the oldest attested method of making fire is by percussion or friction of stone on stone (e.g. flint on pyrite), while friction of wood on wood as employed for example by the Inuit seem to be a later custom (Stapert and Johansen 1999).

Following this theoretical reconstruction of the Lollikhuse dwelling and hearth features, hearth A, found right on your right hand when entering the dwelling was the source for light and heating of boiling stones. After removal of charcoal and shutting of the draft from the door and a probable second opening, also heating by convection could take place. Broiling, grilling and roasting could be done at a hearth of this type, but hearth B is a more effective hearth for these processes. It is not clear whether feature C has a combustion area. More probably this is a storage area and the fire-cracked rocks, ash and bits

of charcoal found here derive from combustion processes in hearth A.

In this article I have suggested a methodological approach to the study of hearth features in general and three examples of reconstructions of the pyro-technology of the hearths of three Stone Age dwellings have been presented. By this I hope to have illustrated the potential of describing hearths as structures composed by elements, which can be described separately allowing their relational function to be judged upon.

Bibliography

BATDORF, C.
1990 *Northwest Native Harvest,* Canada.
BRINK, J. W. and B. DAWE
2003 Hot Rocks as Scarce Resources: The Use, Re-Use and Abandonment of Heating Stones at Head-Smashed-In Buffalo Jump. *Plains Anthropologist* 48 (186): 85-104.
BUCKLEY, V. M.
1990 Experiments using a reconstructed fulacht with a variety of rock types: implications for the petromorphology of fulachta fiadh. In: Buckley, V. (ed.) *Burnt Offerings. International Contributions to Burnt Mound Archaeology*, pp. 170-172.
CLARKE, R. (ed.)
1985 Preface to *Wood-Stove Dissemination. Proceedings of the Conference held at Wolfheze, The Netherlands.* London: Intermediate Technology Publications.
COUDRET, P., M. LARRIERE and B. VALENTIN
1989 Comparer des foyers: une entreprise difficile. In: Olive, M. and Y. Taborin (eds.) *Nature et fonction des foyers préhistoriques. Mémoires du Musée de Préhistoire d'Ile de France* n° 2: 37-46. Nemours.
FAEGRE, T.
1979 *Tents: architecture of the nomads.* London: Murray.
GRØNNOW, B.
1999 Qalunatalik/Polaris site. In: Appelt, M. and H.C. Gulløv (eds.) *Late Dorset in high arctic Greenland: final report on the Gateway to Greenland project.* Copenhagen: Danish Polar Center Publication No. 7. Danish Polar Center: 42-62
JENSEN, J. F.
1998 Dorset dwellings in West Greenland. *Acta Borealia* 15: 59-80, Oslo.
LAUBIN, R. and G. LAUBIN
1989 *The Indian Tipi. Its history, construction and use.* Norman.
MARKSTRÖM, M.
1996 *Skärvsten – vad är det? En experimentell studie.* CD – uppsats i Arkeologi, Vt 1996. Umeå Universitet, Institutionen för arkeologi.

MAXWELL, M. S.
1985 *Prehistory of Eastern Arctic.* New York.
MØBJERG, T. and A.-B. GOTFREDSEN
2004 Nipisat – a Saqqaq Culture Site in Sisimiut, Central West Greenland. Meddelelser om Grønland, *Man & Society* 31.
ODGAARD, U.
1995 *Telte i arktiske miljøer. Rekonstruktioner og ideologi.* Cand. phil. thesis, Institut for Arkæologi og Etnologi, København.
2001 *Ildstedet som livscentrum. Aspekter af arktiske ildsteders funktion og ideologi.* Ph.d. thesis, Forhistorisk Arkæologi, Moesgård, Århus.
2003 Hearth and Home of the Palaeo-eskimos. *Inuit Studies* 27 (1-2): 349-374.
2005 The most extreme Situation. Contextual experiment with an Arctic hearth performed at Lejre Experimental Center. In: *Experimental Pyrotechnology group Newsletter* no. 2, National University of Arts in Bucharest.
OLIVE, M. and Y. TABURIN, (eds.)
1989 *Nature et fonction des foyers préhistoriques. Actes du Colloque International de Nemours 1987. Mémoires du Musée de Préhistoire d'Ile de France* no 2.
OLSEN, B.
1998 Saqqaq housing and settlement in southern Disko Bay, West Greenland. *Acta Borealia* 15: 81-128, Oslo.
PERLÈS, C.
1977 *Prehistoire du feu.* Masson.
SCHLEDERMANN, P.
1990 *Crossroads to Greenland. 3000 Years of Prehistory in the Eastern High Arctic.* The Arctic Institute of North America.
SIRELIUS, U. T.
1906 Über die primitiven wohnungen der finnischen und ob-ugrischen völker. *Finnisch-Ugrische Forschungen* 6: 74-104, 121-154.
SOFFER, O.
1985 *The Upper Paleolithic of the Central Russian Plain.* Orlando: Academic Press.
SOFFER, O., K. VANDIVER and S. VANDIVER
1993 The Pyrotechnology of Performance Art: Moravian Venuses and Wolverines. In: Knecht, H., A. Pike-Tay and R. White (eds.) *Before Lascaux. The Complex Record of the Early Upper Paleolithic.* pp. 259-275, Florida.
STAPERT, D. and L. JOHANSEN
1999 Flint and pyrite: making fire in the Stone Age. *Antiquity* 73 (282), December 1999.
STEFANSSON, V.
1922 *Hunters of the Great North.* New York.
STEWART, H.
1977 *Indian Fishing. Early Methods on the Northwest Coast.* Vancouver.
STEWART, B. *et al.*
1987 *Improved Wood, Waste and Charcoal burning Stoves. A practitioners' manual.* London: IT Publications.
SØRENSEN, S.
1993a Lollikhuse – a Dwelling Site under a Kitchen Midden. *Journal of Danish Archaeology* 11: 19-29.
1993b Lollikhuse – en køkkenmødding ved Selsø. *Arkæologi i Frederiksborg Amt 1983-1993.* Frederiksborg Amt 1993
WATTEZ, J.
1988 Contribution à la connaissance des foyers préhistoriques par l'étude des cendres. *Bulletin de la Société Prehistorique Française*, 85 (10-12): 352-366.

A Social Instrument: Examining the *Chaîne Opératoire* of the Hearth

Silje Evjenth Bentsen

Introduction

The hearth has for a long time been recognized as an important part of site structure (e.g. Binford 1983; Leroi-Gourhan and Brézillion 1966, 1972; *cf.* Gamble 1991). Researchers have focused on the hearth as a centre of clusters of various materials and thus as a centre of different activities. The hearth is, however, in itself an instrument, and an important part of prehistoric technology. It provided heat when cooking and light when knapping flint and it may have had its own functions and meanings to the people using it. The hearth was created and recreated through everyday activities. The hearth was also a social instrument, and knowledgeable actors would use this instrument when producing and reproducing social relations (*cf.* Giddens 1984).

The *chaîne opératoire*, or operational sequences, describes "the series of operations involved in any transformation of matter (including our own body) by human beings" (Lemonnier 1992: 26). The prehistoric hearth has previously been described through operational sequences, focusing on the fitting-out, function and use of the hearth (Taborin 1989). The *chaîne opératoire* will be used to examine the prehistoric hunter-gatherer hearth as an instrument and a phenomenon. This approach will contribute to the understanding of the hearth as an instrument in prehistoric life.

The *chaîne opératoire* of the hearth

Even our most casual acts, like running or eating, are carried out using culturally determined techniques. A technique may be defined as an effective and traditional action that may seem mechanical to the actor (Mauss 1950). A technique consists of tools and gestures, organized in a sequence that is both fixed and flexible (Leroi-Gourhan 1964: 164). Technique and technology are social phenomena, organized in operational sequences. In any operational sequence, there are operations that cannot be replaced, cancelled or delayed without serious consequences on the process and the final result (Lemonnier 1992: 17-24). Different actions might thus consist of different sequences, and slightly different sequences might lead to the same result. There are however certain operations that have to be executed at a certain time to gain a certain result.

The *chaîne opératoire* may be considered both as a conceptual framework and an analytic methodology useful in detailing artefact life-histories. Through the operational sequences of a phenomenon, one gains insight into the decision-making sequences of an activity. This insight may also be used in understanding the social context (Dobres 2000: 164). The approach is well known as part of lithic studies, but has also been used in other contexts. It has, in example, proved useful in studying Aurignacian I personal ornaments and Gravettian anthropomorphic figurines (White 1997).

The hearth is a feature and part of the material culture. It is in itself an assemblage of objects, artefacts, operations and activities (*cf.* Taborin 1989: 78). The *chaîne opératoire* may thus be used to study the various operations of the hearth. The operational sequences of the hearth will give detailed information on the hearth and its life-history.

There are many possible operations associated with the hearth, and there is only room to mention a few of these. I have divided the *chaîne opératoire* of the heart into four stages (*cf.* Bentsen 2005), namely the preparing and the localization of the hearth, the use of the hearth, the reuse of the hearth, and the abandoned hearth. In each of these four stages, I will describe the most important operations of the hearth.

Stage 1: The preparing and localization of the hearth

The first stage in the *chaîne opératoire* of the hearth is the preparing and localization of the hearth. This stage involves operations that have to be executed before the fire is lit, and cancelling or delaying any of these operations will have impact on the hearth and its organization.

An important part of preparing the hearth is the gathering of firewood. This is on the one hand a continuous operation that needs to be performed every day, but it is also an operation that has to be executed in order to use a hearth. When choosing firewood, one has to consider the properties of a material. The calorific value, flammability, persistence and height of the flames and expected duration of heat are important properties of a material (Théry-Parisot 2001: 14-19). Density and moisture are among other significant characteristics (Collina-Girard 1998: 74-81). By analyzing the ashes, charcoal and other remains in the hearth, one might gain information on the type of firewood used and its condition (Théry-Parisot 2001; Wattez 1988, 1991).

Different types of firewood have different properties and might be used in different contexts. When selecting fuel for smoking meat, in example, one might prefer other properties than when collecting wood for a campfire. The wood species available is determined by the environment. Within this range of species, actors will have the possibility to choose. The choice of firewood might thus indicate the level of knowledge of the surrounding environment and the properties of the species available. It may also indicate social practices, in example social rules attached to the use of certain species of wood.

Wood is not the only possible fuel. Bone is also well suitable. Using 20% wood and 80% bone will, on average, keep the fire going twice as long as using 20% bone and 80% wood (Théry-Parisot 2001: Fig. 27). Throughout the Palaeolithic, bones seem to have been the preferred fuel (Théry-Parisot 2001, 116). Fire may have had a symbolic or ritual meaning in the Palaeolithic, in example when used in cremation (Perlès 1977: 150-151). Rituals, ideology or cosmology may also have had significance for the choice of fuel (Théry-Parisot 2001: 14-15). The use of bone as fuel during the Palaeolithic may thus have had a significance beyond the need of heat and light. On the other hand, bone would have the advantage of being easy to come by as it would be a natural by-product from hunting. The hearth may also have been regarded as a place to throw waste (Perlès 1977), and one might not have differentiated between fuel and waste to the same extent as we do today.

The localization of a hearth at a site may have followed various rules and had different explanations. First and foremost, one had to consider the topographic features at the site: The hearth need to be placed at a level surface that is neither too wet nor too dry. The expected function of the hearth would have been significant when placing it. A hearth that is mainly used to dry meat may in example be localized at a different spot than a hearth used for cooking or as a source of light and heat (*cf.* Binford 1978). The localization of the hearth may be based on merely practical or functional reasons, but also on social rules. Certain activities or people might, in example, not be allowed at the same hearths. The size of a hearth may have influence on its localization, but size may again depend on function. In several camps, one would have needed more than one hearth. The localization of one hearth would have influenced the localization of other hearths. In example, there might be social rules connected to the distance between social units (e.g. Binford 1978; 1983).

The placing of a hearth also involves the construction of the hearth itself. The hearth might be lined with stones. Just as one may prefer certain species of firewood, one may also prefer certain species of rock. Rocks have different properties, as evidenced in experiments in replication of fire-cracked rock. In example, hard sandstone seemed to boil water longer and fragment less readily than chert and quartzite. This indicates that hard sandstone would have to be replaced less often than chert or quartzite (House and Smith 1975: 78-79). Size would be another important factor when choosing stones to line a hearth, as would the shape of the rock.

Some hearths are lined with stones, others are not. This may indicate the position of a hearth. In example, a lining of stone may be preferred indoors to prevent dragging out ashes. Outdoor hearths may lack this lining of stones (Binford 1983). The use of stones may also be related to the function of the hearth and the use of the site. One may, in example, choose not to invest too much in a hearth that is expected to be used only once or twice. A hearth that is to be used often at a habitation camp may, however, require a stone lining to keep it contained.

Stage 2: The use of the hearth

When the hearth is prepared and localized, one starts using it. The lightening of the fire is, in most cases, an essential operation when using a hearth. There are many different techniques for lightening a fire. They are divided into friction techniques and striking techniques (Collina-Girard 1998; Hough 1926; Perlès 1977). Striking a piece of flint or quartz against a piece of pyrite is an example of a striking technique. This technique was used already 17000 years ago, and may have been used even earlier. However, pyrite oxidizes easily, and is difficult to trace archaeologically. The technique was still used in the 19th century by certain North American hunter-gatherer bands (Collina-Girard 1998). Friction techniques are various techniques where pieces of wood are rubbed against one another, in example by circular movements (*cf.* Perlès 1977: Figs. 7, 8, 9 and 19).

Coal or ember might be transported and kept over night (Hough 1926). A group may also have practical or ideological reasons for maintaining a constant fire (*cf.* Perlès 1977: 31). The lightening of the fire may thus be an operation that is executed several times, but it may also be an operation that is rarely performed.

The hearth might be part of several activities, and a hearth might have had several functions. Some actors use the hearth as a tool. The hearth might have been the only source of light. It is a source of heat in itself, and one may use it to heat stones and store heat (Perlès 1977). The fire in hearth or the hot stones may be tools for processing meat and cooking. The smoking of meat and processing of ochre are other examples of activities where the hearth is used as a tool. Fire and hearths may also be considered a tool when used as weapons, in example to gain control of caves, keeping animals outside the camp or forcing animals over cliffs (Perlès 1977).

A hearth may function as a meeting point. As a meeting point, the hearth may be the centre of several activities that leave few or no traces. One may in example gather around a hearth to play games or to have discussions.

The hearth may also have a ritual significance when used in cremations or when objects are placed in the hearth, and it may function as a symbol. A domestic hearth may in example serve as a symbol of the household and the unit. A hearth connected to a certain activity or certain functions may also symbolize a specific status or social practice. In example, persons may be placed at specific or separate hearths during transition rites.

The artefacts around the hearth are important in understanding the use of the hearth. A use wear analysis of flint and bone may reveal important information on the activities by the hearth. Seeds and plant remains may show what kind of fuel one used and what kind of plants that were processed by the hearth. Analyses of phosphate are well known indicators of activities, but modern activities have great impact on these kinds of analyzes (Audouze 1989). The size and position of the body are important to the localization of activity areas by the hearth. Waste from activities will make a distinct pattern. Small pieces of bone or flint may create a drop zone around the hearth, whereas large pieces are tossed away and make a toss zone. Larger collections of waste are removed to small dumps at the outskirt of the camp (Binford 1983). Long-term use of the hearth may create one more zone. Small objects, that originally were left where they fell, may be moved to the outskirt of the activity area to avoid accumulations of waste. This will create a displacement zone between the drop zone and the toss zone (Stevenson 1985; 1991).

The contents of the hearth may also give information on the function of the hearth and the hearth as an instrument. The ashes and charcoal may give information on the fuel, and pieces of fire-cracked rock may point to the heating of stones for cooking. Ritual deposition of objects in or by the hearth may also be found archaeologically. In example, one of the stones lining a hearth at the Magdalenian site Étiolles was engraved. The stone was probably not placed there while the fire was lit, as the heat and the fire would have altered the engraved figures (Taborin *et al.* 2001). This indicates that the hearth also had functions and was used as an instrument independent of the fire.

Stage 3: The reuse of the hearth

The reuse of the hearth may be a challenge to the archaeologist. It may be difficult to find the difference between a hearth where many actors have executed one activity and a hearth where the same activity was executed numerous times by one or a few actors. It is thus important to know the operational sequences of different activities in order to separate activities and operations. Some operations of an activity might not be executed by the hearth or the hearth and its surroundings might be cleaned. As a result, certain operations are not present by the hearth.

The hearth itself may also be influenced by cleaning. In example, one might get a hollow shape if the hearth was cleaned out several times. Reuse may also change the size and shape of the hearth. In example, a change of function may lead to the need of a larger or smaller hearth.

All types of rock change under thermal influence, and might get a different colour, become lighter and shrink. Heat also makes rocks crack unevenly, they become more easily to shape, rocks might crackle, become porous or granular, and heat might create pot lids. How a certain stone changes depends on the type of rock, the heat, the time of the thermal treatment and the speed of cooling (House and Smith 1975; Luedtke 1992). Rocks may in addition be chemically altered under thermal influence, depending on the type of rock and the type of subsoil (Luedtke 1992). Heat will also affect bones. In example, they may shrink and the colour might change (Shipman *et al.* 1984 in Reitz and Wing 2001). By examining the bones and stones in and around the hearth, one may draw conclusions on the use and reuse of the hearth. In example, on might examine the stones in the lining. If stones of the same type of rock show the same signs of thermal changes, this indicates that they have been in the hearth the same amount of time. If not, the hearth might have been rearranged or rocks replaced before the reuse of the hearth.

The subsoil may change colour when heated, and this might indicate the length and amount of heat (Wattez 1991: 5). The subsoil and earth might also be analyzed to examine the use and reuse of the hearth. A large amount of seeds and plant species might indicate long term use or reuse. The analyses of the sediments in the subsoil and the stratification might also indicate if the hearth was continually used or abandoned before reused (Wattez 1991). One may also do chemical analyses of the ashes from the hearth. Ashes are made by carbonate crystals. These crystals have different shapes and sizes, depending on the conditions when they are shaped and the development of these conditions (Courty 1984 in Wattez 1988). Ashes thus reveal information on the intensity of fire and the use of the hearth. However, the most intense heating will always have the largest impact on the hearth.

Stage 4: The abandoned hearth

Hearths are abandoned when they are of no use. This is the stage of the archaeologists: We work with the abandoned hearth (*cf.* Coudret *et al.* 1989: 38; Perlès 1977: 45). A hearth might be cleaned and removed, and its remains found in dumps. One might also make a dump at a previous hearth. The hearths that are used continually while staying at the site are abandoned when the camp is left. We do not necessarily find remains of all stages and operations in a hearth, and may have to search the dumps to find information on the *chaîne opératoire* of the hearth.

The people that inhabited a site, and used the hearth one last time, knew they were going to leave. If they expected to use the camp and the hearth again, this may have influenced how they left the site. In example, the *Nunamiut* abandon seasonal camps with useable gear and features, like small storages of raw materials for tools, hearths and firewood (Binford 1978). If people at the site did not expect to use the camp again, they may on the other not have cared how the hearth was left. In example, one might have avoided cleaning the hearth. The abandoned hearth may thus not correspond to the functional condition of a hearth (Coudret *et al.* 1989: 43).

An archaeological material is influenced by post-depositional processes. These are divided into cultural and non-cultural (or natural) processes (Schiffer 1972). The cleaning and reuse of the hearth would be cultural processes, while natural processes might be started by flooding, wind, animals and chemical processes (Hofman 1986; Schiffer 1983). Natural post-depositional processes might move charcoal and influence the chronology of the hearth (Théry-Parisot 2001; Wattez 1988). Reuse might happen after a long time, in example when other cultural groups start using a hearth or settles at or by an abandoned camp. This is a prehistoric post-depositional process, whereas the digging of drainage trenches is a fairly modern cultural process. The archaeological excavation is also a cultural post-depositional process, with permanent consequences for the hearth. Post-depositional processes will influence the knowledge one might get from a hearth. In example, reuse might have consequences for the dating and the appearance of the hearth, while the digging of drainage trenches may destroy a hearth completely.

The instrument

To the archaeologist, a hearth is a feature. A hearth is however also, as shown above, the end product of a process consisting of several operations. The people that inhabited a site chose bone or firewood for the fire and gathered rocks to line their hearths. The choice of wood species and type of rock may are operations that indicate the actors' knowledge of their environment. These operations may also indicate cultural preferences, like preferences relating to the type or size of the rocks used by the hearth.

The hearth was used in many activities. It was a source of heat and light, and functioned as a tool in several activities. It would in example have been a tool when cooking and processing ochre. The hearth may have been used as a weapon in hunting or to keep animals outside the camp. It may also have functioned as a meeting point and a symbol, and it may have had ritual significance.

The hearth was an instrument that was used and reused daily, and several analyzes may be used to gain information on the use and reuse of the hearth.

The social instrument

The hearth was an important instrument in everyday life in prehistory and a feature every actor needed. The hearth is recognized as important to the spatial organization at a site. Spatial organization may have been governed by differentiation relating to gender, age and status (*cf.* Bourdieu 1990). In other words, actors may position themselves in time and space and relationally (Giddens 1984). A social position is "*a social identity that carries with it a certain range (however diffusely specified) of prerogatives and obligations that an actor who is accorded that identity (or is an 'incumbent' of that position) may activate or carry out; these prerogatives and obligations constitute the role-prescriptions associated with that position*" (Giddens 1979: 117).

There are many examples of how the social position of an actor may have influenced his or her position by the hearth. The hearth may in example have had a male and a female side (e.g. Grøn 1991). Skills may have been important to the status of the actor and may also have determined the seating by a hearth. It has been suggested that skilled flintknappers at habitation U5 at the Magdalenian site Étiolles sat close to the hearth. Unskilled knappers may have been placed along the edges of the habitation, indicating that children and novices would practice by the sleeping areas, where they would not hinder economic activities (Pigeot 1990: 132; *cf.* Pigeot 1987). Certain activities, actors and/or positions may also have been attached to certain hearths, and this may have had impact on the positions at the hearth and the localization of other hearths at a site. If, in example, a hearth represents a household, other hearths and households may be placed at a given distance (e.g. Binford 1978; 1983).

The use of the hearth and the positioning by the hearth would have been a routinized practice. The routinization of practice is essential to the actors, as a sense of trust and security is grounded in the routines of day-to-day life. People draw upon structure, or rules and resources, in their day-to-day activities. Structures are organized as properties of social systems, and are both medium and outcome of social practices. They exist as memory traces and in their instantiations in social practices, and are both enabling and constraining. Agent and structure thus represent a duality in a social system (Giddens 1984).

The prehistoric hearth was continually created and recreated by the actors. The *chaîne opératoire* of the hearth show insight into the social practices at a site. The localization of a hearth may, to begin with, indicate its function(s) and/or status. The relation and distance to other hearths at the site may, in example, be grounded in the social practices of spatial organization and the distance between different households and units.

Secondly, the social practices of the use of the hearth are shown through the activities by and operations of the hearth. There may have been rules governing the use of the hearth and the kinds of activities that could be performed at certain hearths. Some hearths may in example have been used merely as tools, while some had a ritual significance and yet others had several functions.

The positions of the actors by the hearth may also reflect their social positions and the social structure. If the hearth was divided in a male and female side, one may in example expect different material patterns at the different sides of the hearth. If a hearth, on the other hand, was reserved for a certain actor or group of actors, the artefacts around the hearth may reflect this. The hearth itself may also stand out if it served as a symbol of special social positions, in example it may have been lined with a certain kind of stone.

Concluding remarks

The *chaîne opératoire* may be used to gain insight into the hearth. There are many operations relating to the hearth, and a selection of operations has been presented. The operational sequences of the hearth were divided into four stages; the preparing and localization of the hearth, the use of the hearth, the reuse of the hearth, and the abandoned hearth.

The hearth is an end product of many activities. The hearth was an instrument in prehistoric life, and the operations that created and recreated the hearth give information on the use of this instrument. It may also give information on the social practices of the people that inhabited a site. While creating the hearth and working by it, the people would create and recreate their social practices. The archaeologists may today use the material culture, including the hearth, as an instrument in understanding the people that lived in the camp.

Acknowledgements

I would like to thank for the invitation to contribute to this book. Thanks to Dr. Sheila Coulson at the University of Oslo for supporting my work and to the many anonymous helpers who commented an earlier draft of this paper. Needless to say, all shortcomings of this paper are the responsibility of the author.

Bibliography

AUDOUZE, F.
  1989 Le dosage des phosphates. *Le Courier du CNRS* 73: 15.

BENTSEN, S. E.
  2005 *Individet og ildstedet: Sosial struktur i seinpaleolitikum.* (Translated: *The individual and the hearth: Social structure in the Upper Palaeolithic.*) University of Oslo: Unpublished M.A. thesis.

BINFORD, L. R.
  1978 *Nunamiut Ethnoarchaeology.* New York: Academic Press.
  1983 *In Pursuit of the Past.* London: Thames and Hudson.

BOURDIEU, P.
  1990 *The Logic of Practice.* Cambridge: Polity Press.

COLLINA-GIRARD, J.
  1998 *Le Feu avant les Alumettes: Expérimentation et Mythes Téchniques.* Collection Archéologie Expérimentale et Ethnographie des Techniques. Paris: MSH.

COUDRET, P., M. LARRIÉRE and B. VALENTIN
  1989 Comparer des foyers: une entreprise difficile. In M. Olive and Y. Taborin (eds.) *Nature et Fonction des Foyers Préhistoriques: Actes du Colloque International de Nemours 12-13-14 mai 1987.* Mémoires du Musée de Préhistoire d'Île de France 2. Nemours: A.P.R.A.I.F, pp. 37-45.

COURTY, M. A.
  1984 Formation et évolution des accumulations cendreuses. Approche micromorphologique. *Actes du Colloque Interrégional sur le Néolithique, le Puy-en-Velay*, 3 et 4 octobre 1981. France: CREPA, pp. 341- 353.

DOBRES, M.-A.
  2000 *Technology and Social Agency: Outlining a Practice Framework for Archaeology.* Oxford and Malden: Blackwell Publishers Ltd

GAMBLE, C. S.
  1991 An introduction to the living spaces of mobile peoples. In *Ethnoarchaeological Approaches to Mobile Campsites.* Ethnoarchaeological Series 1. Ann Arbor: International Monographs in Prehistory.

GIDDENS, A.
  1979 *Central Problems in Social Theory: Action, Structure and Contradiction in Social Analysis.* London: Macmillan Press.
  1984 *The Constitution of Society.* Cambridge: Polity Press.

GRØN, O.
  1991 A method for reconstruction of social structure in prehistoric societies and examples of practical application. In O. Grøn, E. Engelstad and I. Lindblom (eds.) *Social space.* Odense University Studies in History and Social Sciences 147. Odense: Odense University Press, pp. 100-117.

HOFMAN, J. L.
  1986 Vertical Movement of Artefacts in Alluvial and Stratified Deposits. *Current Anthropology* 27: 163–171.

HOUGH, W.
1926 *Fire as an Agent in Human Culture.* Smithsonian Institution United States National Museum Bulletin 139. Washington: Smithsonian Institution.

HOUSE, J. H. and J. W. SMITH
1975 Experiments in replication of fire-cracked rock. In Schiffer M. B. and J. H. House (eds.) *The Cache River Project: An Experiment in Contract Archaeology.* Arkansas Archaeological Survey, Publications in Archaeology, Research Series No. 8. Arkansas: University of Arkansas Press, pp.75–80.

LEMONNIER, P.
1992 *Elements for an Anthropology of Technology.* Ann Arbor: Anthropological Papers, University of Michigan No. 88.

LEROI-GOURHAN, A.
1964 *Le Geste et la Parole: I – Téchnique et langage.* Paris: Albin Michel.

LEROI-GOURHAN, A. and M. BRÉZILLION
1966 L'habitation magdalénienne no 1 de Pincevent près de Montereau (Seine-et-Marne). *Gallia Préhistoire* 9: 263-371.
1972 Fouilles de Pincevent, Essai d'Analyse Ethnographique d'un Habitat Magdalénien. *Gallia Préhistoire*, Supplément 7.

LUEDTKE, B.
1992 *An Archaeologist's Guide to Chert and Flint.* Archaeological Research Tools 7. Los Angeles: University of California Press.

MAUSS, M.
1950 *Sociologie et anthropologie.* Paris: Presses Universitaires de France.

PERLÈS, C.
1977 *Préhistoire du feu.* Paris: Masson.

PIGEOT, N.
1987 *Magdaléniens d'Étiolles: Économie de Débitage et Organisation Sociale (l'Unité d'Habitation U5).* Gallia Préhistoire, Supplément 25. Paris: CNRS.
1990 Technical and social actors. Flintknapping specialists and apprentices at Magdalenian Étiolles. *Archaeological Review from Cambridge* 9(1): 126-141

REITZ, E. J. and E. S. WING
2001 *Zooarchaeology.* Cambridge manuals in Archaeology. Cambridge: University of Cambridge.

SCHIFFER, M. B.
1972 Archaeological Context and Systemic Context. *American Antiquity* 37: 156–165.

1983 Toward the Identification of Formation Processes. *American Antiquity* 48: 675–706.

SHIPMAN, P., G. FOSTER and M. SCHOENINGER
1984 Burnt bones and teeth: An experimental study of color, morphology, crystal structure and shrinkage. *Journal of Archaeological Science* 11: 307-325.

STEVENSON, M. G.
1985 The Formation of Artifact Assemblages at Workshop/Habitation Sites: models from Peace Point in Northern Alberta. *American Antiquity* 50: 63–81.
1991 Beyond the Formation of Hearth-Associated Artifact Assemblages. In Kroll E. M. and T. D. Price (eds.) *The Interpretation of Archaeological Spatial Patterning: Interdisciplinary Contributions to Archaeology.* New York and London: Plenum Press, pp. 269–296.

TABORIN, Y.
1989 Le foyer: Document et concept. In Olive M. and Y. Taborin (eds.) *Nature et Fonction des Foyers Préhistoriques: Actes du Colloque International de Nemours 12-13-14 mai 1987.* Mémoires du Musée de Préhistoire d'Île de France 2. Nemours: A.P.R.A.I.F, pp. 77-80.

TABORIN, Y., M. CHRISTENSEN, M. OLIVE, N. PIGEOT, F. CAROLE and G. TOSELLO
2001 De l'art Magdalénien figuratif à Étiolles (Essonne, Bassin Parisien). *Bulletin de la Société Préhistorique Française* 98: 125–128.

THÉRY-PARISOT, I.
2001 *Économie des Combustibles au Paléolithique.* Dossier de documentation archéologique n° 20. Paris: CNRS.

WATTEZ, J.
1988 Contribution à la connaissance des foyers préhistoriques par l'étude des cendres. *Bulletin de la Société Préhistorique Française* 85: 352-366.
1991 *Dynamique de Formation des Structures de Combustion de la Fin du Paléolithique au Néolithique Moyen. Approche Méthodologique et Implications Culturelles.* University of Paris I: Unpublished Ph.D. Thesis.

WHITE, R. R.
1997 Substantial acts: From materials to meaning in Upper Palaeolithic representation. In Conkey, M., O. Soffer, D. Stratmann and N. G. Jablonski (eds.) *Beyond Art: Pleistocene Image and Symbol.* San Francisco: University of California Press, pp. 93-121.

Etude du profil thermique d'une structure de combustion en meule (pitkiln): four ou foyer simple?

Claude Sestier

Résumé

Le profil thermique d'une structure de combustion de type meule (*pit-kiln*) a été étudié au cours d'une série d'expériences, en mesurant le gradient de chauffe (°C/mn) des poteries, la durée de chauffe supérieure à 750°C et le refroidissement des poteries, au cours de cuisson de poteries de composition et dimension variées. L'influence de plusieurs paramètres sur le profil thermique est décrite. Ce travail confirme que ces paramètres sont utiles pour décrire le fonctionnement de cette structure non permanente. Le fonctionnement de cette structure «en meule» est similaire à celui d'un four.
Mots-clefs: meule, profil thermique, technologie, poterie.

Abstract

The thermal profile of a pit-kiln has been studied by measuring the heating rate of the fired-pots, their soaking time (over 750°C) and their cooling rate. This work points out that the function and efficiency of a firing structure cannot be always infered from its architecture and organisation, especially when the remains are poorly organized.
Keywords: pit kiln, thermal profile, technology, potery.

Introduction

Les techniques de cuisson des poteries et les structures de cuisson: Les techniques de cuisson des poteries archéologiques antéhistoriques ont fait l'objet de nombreux travaux, tant concernant les structures de combustion (Drews 1978, Duhamel 1978, Petrasch 1986, Lüdtke et Vossen 1991) que l'effet de la température sur la matrice argileuse et les éléments qui l'accompagnent (Shepard 1956, Rye 1981, Gibson et Wood 1997, Rice 1987, Arnal 1989, Orton *et al.* 1993).

Sur ce sujet, on peut schématiquement organiser la littérature archéologique et ethnologique en trois thèmes: les contraintes techniques analysables selon des principes physiques, les choix techniques culturellement définis, la valeur économique et sociale de la poterie.

On s'intéressera en particulier au premier thème, car il permet une observation directe soit de pratiques actuelles, soit de pratiques reconstituées dans un cadre visant à caractériser objectivement une pratique technique et tester des hypothèses. Une question importante pour les archéologues est de savoir quelle relation existe entre la qualité de cuisson des poteries (démontrable selon certains critères) et certaines structures de combustion dont il s'agit d'évaluer l'aptitude à la production de poteries.

Caractérisation du fonctionnement des structures de combustions: Il est généralement admis qu'il existe une relation entre le degré de complexité de la structure de combustion et le degré de savoir-faire et d'évolution de la pratique potière. Le fonctionnement des structures de cuisson peut être interprété *a priori* selon des principes physiques simples (Kingery 1997). La fermeture de la structure de combustion permet tout d'abord de mieux utiliser la chaleur puisque les gaz chauds doivent circuler entre les poteries. La cuisson en meule ou en fosse présente donc cet avantage, en comparaison à une cuisson en foyer totalement ouvert.

Une séparation de la zone de combustion et de la zone de cuisson permet un apport de combustible et de comburant (l'oxygène de l'air) indépendant de l'organisation des objets dans la zone de cuisson. Ceci permet de faire des cuissons plus longues, ou de modifier l'atmosphère de cuisson. Le contrôle du flux thermique est plus aisé, et on obtient une plus grande homogénéité des températures dans la zone de cuisson, en raison d'une circulation forcée entre les poteries.

Dans tous les cas, la cuisson des matériaux argileux nécessite d'atteindre des températures supérieures à 700°C, une valeur seuil qui dépend de la composition minéralogique des argiles.

On parlera de céramique (et non plus de poterie) quand apparaitront certaines phases minéralogiques, produit d'un véritable métamorphisme. En fait, cette transition de la poterie à la céramique dépend des températures atteintes, des durées d'exposition à la chaleur (le phénomène suivant une cinétique chimique qui lui est propre), et de l'atmosphère de cuisson. Ceci conduit à parler d'équivalent-température de cuisson (*equivalent-firing temperature*), des cinétiques différentes pouvant donner le même résultat final sur le tesson.

Les structures de cuisson de poterie sont-elles auto-explicatives? : Ce n'est que pour des périodes très récentes que les structures de combustion portent des caractéristiques structurelles assez explicites pour qu'on puisse en inférer un mode de fonctionnement spécifique (Vertet 1978), et qu'on puisse assurer qu'une volonté d'optimisation du processus a été suivie. Le terme de four (*kiln*) est donc réservé à une structure permanente, ménageant un espace relativement clos, avec une ouverture à la base et un orifice supérieur permettant l'évacuation par convection des gaz de combustion. Certains auteurs donnent une définition plus restrictive, le four devant permettre une séparation complète entre les poteries et le combustible (Balkansky *et al.* 1997). On oppose donc classiquement à ces fours des procédés considérés comme plus simples et/ou moins évolués, comme la cuisson en foyer ouvert type «feu de camp» (*bonfire*), la cuisson en meule (*pit-kiln*) et la cuisson en fosse, une variante étant une fosse à paroi verticale (*updraught-kiln*). Les structures permanentes qui y sont associées sont généralement très simples, voire inexistantes (soit parce qu'elles n'existent pas, soit parce qu'elles sont détruites à la fin de la cuisson).

Il apparaît qu'il n'est pas possible d'opposer de façon simpliste diverses structures de combustion uniquement sur la base de leurs caractères permanents comme leur architecture, et qu'il faut prendre en compte le savoir-faire du potier pour ce qui concerne la conduite du feu. Certains moyens parfois jugés comme primitifs par les nations industrielles sont en effet largement utilisés par diverses ethnies et participent à une économie locale très active. L'observation de ces pratiques actuelles, dans différents contextes économiques et culturels, alimente une réflexion sur la technique potière et ses relations avec la société (Arnold 1991, Gosselain et Livingstone-Smith 1997, Gosselain 1992). Enfin et surtout, les observations réalisées par ces équipes remettent en question certaines prémisses d'un raisonnement archéologique classiquement admis.

En effet, on suppose habituellement qu'à une structure de combustion correspond un profil thermique déterminé. Un corollaire est que la température maximale atteinte par les poteries (qu'on suppose déterminable selon certains critères physiques) serait donc liée à une structure de combustion.

En fait, sur la base d'observations ethnologiques, O. Gosselain conclut ainsi: *a) la température maximale de cuisson ne permet pas de différencier les différentes modalités opératoires de cuisson, correspondant à des structures de combustion et des combustibles variés, b) le combustible n'a pas d'influence sur la température maximale atteinte, c) la température maximale obtenue est très variable au sein d'un même foyer, d) la température est un paramètre très variable selon la position dans la structure de combustion et même au sein d'une même poterie, ainsi qu'au cours du temps*

(Gosselain 1992). A. Livingstone-Smith (2001) complète ces observations et synthétise l'observation de 105 profils thermiques de cuissons observées dans différentes ethnies du Cameroun, Togo, Burkina-Faso, et Sénégal (Gosselain et Livingstone-Smith 1997, Livingstone-Smith 2001).

Ces observations ethnologiques contredisent donc un présupposé classique comme l'opposition entre feu ouvert (*open-firing*) et cuisson en four (*kiln-firing*), où on devrait avoir une cuisson rapide dans le premier cas et lente dans le deuxième. Par contre, elle met en avant l'importance de la conduite du feu et la part du savoir-faire dont certaines règles sont culturellement transmises.

A. Livingstone-Smith montre aussi que la diversité des structures de combustion et des protocoles d'utilisation traditionnels donnent des profils thermiques qui peuvent être classés en tenant compte de la durée de leur palier de cuisson (*soaking-time*) et des montées en température (°C/mn).

Nous proposons ici de tester une structure qui ressemble fortement à celle décrite sous le terme de «*pit-kiln*», par exemple au Mexique (Balkansky *et al.* 1997, fig. 12). Ce type de structure impermanente, rencontré à diverses occasions par les archéologues, est à mettre en opposition avec des structures plus permanentes (*updraught kiln*), qui peuvent être présentes dans la même région (Pool 2000), le choix d'un type de structure de cuisson dépendant de divers facteurs autres que techniques.

Lorsque de telles structures sont associées à de la poterie très bien cuite et aux couleurs bien contrôlées, elles posent à l'archéologue un problème puisqu'on considère habituellement qu'une telle production élaborée ne pourrait se faire qu'avec un « four » et non par des moyens aussi primitifs.

Notre hypothèse de travail est qu'une simple structure en «*pit-kiln*» a un fonctionnement analogue à celui d'un four et permet des résultats reproductibles et en grande partie contrôlables.

Nous montrons ici qu'une structure aussi simple permet d'obtenir de façon reproductible des montées en température et des paliers de cuissons > 700 °C, dont la durée peut être contrôlée. On montre aussi qu'une modification simple de la structure peut influencer de façon significative son comportement thermique. Il est possible d'y contrôler l'homogénéité des couleurs de cuisson en protégeant les poteries du contact avec le combustible (Rice 1987: 154).

Matériel et méthodes

Principe de construction de la structure (**Fig. 1**): la structure est composée de trois éléments: les poteries à cuire, le combustible et des pierres de volume variable.

Le combustible est mêlé aux poteries. On compte en général 2 volumes de combustible pour un volume de poterie (estimation visuelle). Cet assemblage forme un dôme qui est progressivement recouvert de pierres de volume moyen (10-20 cm), les interstices étant occupés par de plus petites pierres (4-8 cm).

Généralement, on allume le feu à la base de la structure à moitié constituée et on recouvre totalement les poteries de bois puis de pierres (**Fig. 1C**).

Variation de structure: la structure fait à peu près 60 cm de haut, et contient de 3 à 6 poteries selon le volume disponible. Le diamètre est égal à la hauteur (TYPE 1) ou bien audouble de sa hauteur. (TYPE 2). Le type 1 ou 2 est dit couvert lorsque la totalité de la structure est recouverte de pierres calcaires. Au contraire, le type 1 ou 2 est dit ouvert en l'absence de couverture de pierre (on voit alors le combustible). Dans de rares cas, on a réalisé une chape d'argile sur la structure couverte, en réalisant une fermeture partielle de la structure (**Fig. 1D**), au moment du maximaum de chauffe.

Combustible: on a utilisé du bois mort pris sur pied, soit du petit bois (diamètre < 5 cm), soit du plus gros calibre (diamètre compris entre 15 et 20 cm). Les espèces sont essentiellement du Chataignier (*Castaneus*) et du Hêtre (*Fagus*). La quantité de combustible a été pesée.

Mesure de la température: on a utilisé des thermocouples K préalablement étalonnés. Les sondes sont en contact avec la poterie, soit sur la face externe, soit sur la face interne. Les poteries sont positionnées ouverture vers le sol (**Fig. 2C**), pour faciliter la mise en place des thermocouples, maintenus en contact à l'aide de fil métallique et du poids des pierres. On a aussi réalisé des mesures complémentaires de température, 10 cm en dessous du foyer ou au contact de la partie externe des structures fermées.

Poteries: elles sont réalisées en matériaux variés mais suffisamment poreux pour supporter un dégazage rapide et des gradients de température importants (argiles fortement dégraissées). L'exemple de la figure 1A montre des fac-similés de poteries de style AUVERNIER (Néolithique Suisse) réalisés par R. Martineau en marne fossilifère (O. Acuminata). Elles pèsent 4 et 9 kg.

Résultats

Evolution du feu (**Fig. 1**, structure couverte): on observe 4 phases, une phase de séchage du combustible avec condensation d'eau sur la couverture de pierre, une phase de distillation du bois avec production de fumées colorées puis blanches (**Fig. 1C**), une phase sans production de fumée, où on a transformé la quasi-totalité du combustible en charbon de bois (**Fig. 1D**), une phase de diminution du combustible avec affaissement de la structure (**Fig. 1E**). Dans la pratique, et en dehors de cette expérience, on interrompt la cuisson des poteries bien avant ce stade pour éviter une surcuisson, en particulier pour des argiles à inclusions calcaires pouvant donner de la chaux.

Consommation de bois: le ratio 2 volumes de bois mort pour 1 volume de poterie a permis dans tous les cas une cuisson correcte.

Profil thermique: On peut voir en figure 2A une courbe typique d'évolution de la température, avec trois phases. Une phase de montée en température, une zone de ralentissement (plateau de largeur variable), puis une phase de décroissance de température. Entre 300 et 700°C, on calcule l'accroissement de la température (°C/mn). On note la température maximale, la durée t >750°C pendant laquelle la température se maintient à plus de 750°C, le taux de décroissance de la température sur les 2 heures de décroissance, et après 2 heures (lorsque c'est possible).

On remarque que l'enregistrement de la température sous le foyer (**Fig. 2A**) peut donner une indication fiable de la production de chaleur du foyer, enregistrée ici par un effet de diffusion thermique. On voit de façon très claire la fin de la combustion avec un point d'inflexion correspondant au début de refroidissement.

Le profil thermique en figure 2B est obtenu avec un foyer identique à 2A, sauf que la couverture de pierres a été complétée par une chape d'argile, cependant pas totalement étanche (**Fig. 2D**). On remarque que cela a pour effet de diminuer considérablement le refroidissement du foyer. Les irrégularités du profil de température de la couverture externe d'argile (pointillés sur le graphique) sont dues à des écroulements partiels de la couverture, laissant alors le thermocouple au contact des gaz chauds (2 avec flèche).

Fig. 1: A) Deux pots avant cuisson (fac-similé R. Martineau, style néolithique Auvernier, 4 et 9 Kg en poids sec, Marne à O. Acuminata. B) Montage du foyer, mélange de bois et poteries. C) Phase postérieure au séchage du bois, fumées blanches: début de la montée en température. D) Phase de cuisson sans fumées. E) Ecroulement de la structure. F) Les deux poteries cuites (ici: sur-cuites), avec coloration homogène due à leur isolation du combustible par des petites pierres.

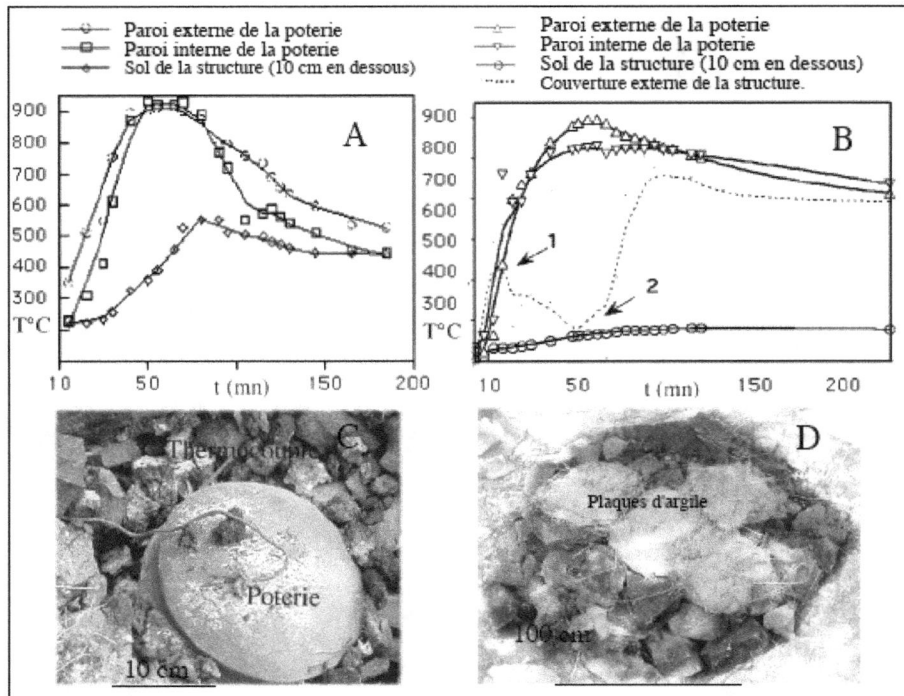

Fig. 2: A) Profil thermique caractéristique d'une structure de type 1 ou 2. B) idem, mais couverture partielle avec des plaques de torchis (2D). C) vue d'un thermocouple en contact avec la partie externe de la poterie, après défournement.

Paramètres thermiques pour 3 types de foyers									
Combustible					Chauffe		Durée plateau		Refroidissement
Structure H=60 cm	petit bois Ø <5 cm P(g)	gros bois Ø 15-20 cm G(g)	Poids total P+G(g)	G/P	°C/mn ±0.2	Tmax °C ±20	Plateau durée en mn >750°C	>900°C	°C/mn (±0.5) 120 mn °C /mn ±(0.2) 120 a 240 mn
TYPE 1 H=D couvert	900	4600	5500	5.1	13	870	100	0	1.5+ 2-4+
	900	4750	5650	5.3	12	900	90	30	2 ND
	1700	5000	6700	2.9	14	980	100	40	2 ND
	900	7400	8300	8.2	13	930	100	25	2 ND
	5000	25000	30200	5	14	980	150	30	2 ND
	600	30400	31000	51	12	910	140	10	2 ND
	650	31200	31850	48	12	960	135	35	1+ 2-3+
	400	31700	32100	79	14	850	150	0	2 ND
TYPE 2 couvert H=D/2	1400	3800	5200	2.7	12	920	20	5	3 ND
	3600	11800	15400	3.3	14	970	30	35	3 ND
	500	30100	30600	60	13	990	90	35	3 ND
	600	33600	34200	56	13	980	110	30	3 ND
TYPE 2 ouvert H=D/2	1600	4800	6400	3	85±5	980	15	6	10-20 ND
	1700	5000	6700	3	85±5	990	10	5	10-20 ND

Fig. 3: Profil thermique pour trois types de foyer. G/p = rapport pondéral gros bois et petit bois, refroidissement: mesuré pour T<700°C.

L'ensemble des 14 expériences est consigné dans les figures 3 et 4. On a fait varier plusieurs paramètres avec des valeurs hautes et basses: combustible (calibre), géométrie de la structure de combustion (rapport H/D), couverture ou non de la structure (une structure non couverte est ici assez proche d'un foyer ouvert et non d'un «*updraught-kiln*»).

Accessoirement, on a regardé l'effet d'un chapage partiel de la structure (**Fig. 2B**). L'ensemble des expériences est conçu pour pouvoir être repris selon une méthodologie des plans d'expérience (Goupy 1985).

Discussion des résultats

Rapport «petit bois/gros bois»: il n'influence pas sensiblement T_{max} et le gradient de chauffe (°C/mn). On ne peut faire ici de commentaires, car il manque certainement des expériences pour vérifier cette absence de relation. La consommation de bois est remarquablement faible pour obtenir une cuisson correcte, (on l'estime au 1:10 de ce qui est souvent décrit dans la littérature «expérimentale»).

Température maximale: T_{max} ne dépend pas de façon significative de la géométrie de la structure de combustion, ni du fait qu'elle soit couverte ou ouverte. En fait, si la cinétique de combustion peut dépendre de la structure de combustion, le T_{max} dépend du pouvoir calorifique du combustible (joules fournies / unité de poids et de volume).

Gradient de chauffe: il est indépendant du rapport H/D, tant que la structure est fermée. Par contre, si la structure est ouverte, le gradient de chauffe est très supérieur (80 à 90°C /mn) avec une durée de chauffe t >750°C très court ainsi qu'un refroidissement 10 fois plus rapide (10 à 20°C/mn au lieu de 2 à 3°C/mn). Ce gradient de chauffe, une fois le combustible totalement sec, résulte d'un équilibre entre la production de chaleur (limitée par l'apport d'oxygène) et les pertes thermiques par convection et rayonnement. On conçoit donc facilement que divers paramètres peuvent intervenir comme l'agencement du combustible, la circulation d'air, etc. Dans l'exemple présenté, le peu de variation observée résulte du peu de variation dans la procédure utilisée pour conduire le feu (montage de la structure, assemblage du bois, etc.).

Effet de la géométrie de la structure sur le profil thermique (**Fig. 4A**: on observe un effet significatif du rapport H/D sur la durée de chauffe <750°C (*soaking time*)). Plus la hauteur est grande par rapport au diamètre, plus les effets de convection sont importants, augmentant certainement le tirage du feu et donc diminuant le « soaking-time ». Le fait que le gradient thermique reste inchangé résulte probablement de la compensation de deux effets antagonistes (plus grande production de chaleur car augmentation du tirage mais plus grande perte

par ce même tirage). Une structure basse permet d'augmenter le «*soaking-time*».

Cet effet interagit positivement avec la quantité de combustible. Dans ce système, plus la masse de combustible est élevée, plus la combustion est longue. Le foyer ouvert obéit à ces mêmes lois, avec une convection et une perte thermique maximales. Ce système est donc prédictible et contrôlable par différents moyens, principalement en réglant le tirage (on peut fermer les orifices entre les pierres), en adaptant la quantité de combustible à la longueur prévue de cuisson, et en stoppant la cuisson.

Il est remarquable que la conduite du feu dans les structures présentées ici ne dépend pas de facteurs météorologiques incontrôlables (vent) ni de la qualité du bois (humidité), ce qui laisse à penser que le rôle régulateur de la structure de combustion est très important.

La même constatation peut être faite pour les résultats présentés par A. Livingstone-Smith (2001): on peut constater que la variabilité des paramètres thermiques est beaucoup moins forte dans les structures de «four» que dans les autres. Si ce dernier insiste sur le fait que «*the structure is by no means the major parameter of firing technologies*», c'est quand même la structure qui autorise une certaine conduite du feu, et donc l'application d'un savoir-faire qui reste cependant essentiel.

Selon les données de Livingstone-Smith (2001), le profil thermique de nos conduites de feu est en fait comparable au profil thermique de four (**Fig. 4C**). Ceci conduit à plusieurs remarques: pour un futur archéologue, une telle structure ne laisse quasiment rien d'organisé, devant être démontée pour récupérer les poteries cuites. Cependant, la qualité des cuissons obtenues est comparable à ce qui pourrait être obtenu dans une structure permanente (four), d'autant que la séparation de la poterie et du combustible par des petites pierres permet d'éviter les taches de carbone et donne une coloration homogène. Si le procédé n'est pas aussi pratique que dans un four avec séparation des poteries et du combustible, il est techniquement valable.

Ce travail confirme que les paramètres thermiques «*heating rate*» et «*soaking time*» doivent être utilisés non pour inférer l'existence de structures de combustion particulières, mais plutôt pour analyser les traditions de cuisson en terme de fonctionnement. La caractérisation physique des poteries en fonction des contextes de production en est le complément indispensable, afin de prendre en compte les qualités et défauts des matériaux argileux (pour éviter par exemple une surcuisson), établir les conditions minimales de cuisson (certains matériaux cuisent plus facilement que d'autres), et intégrer certaines données comme l'économie des combustibles dans le schéma socio-économique de la production potière.

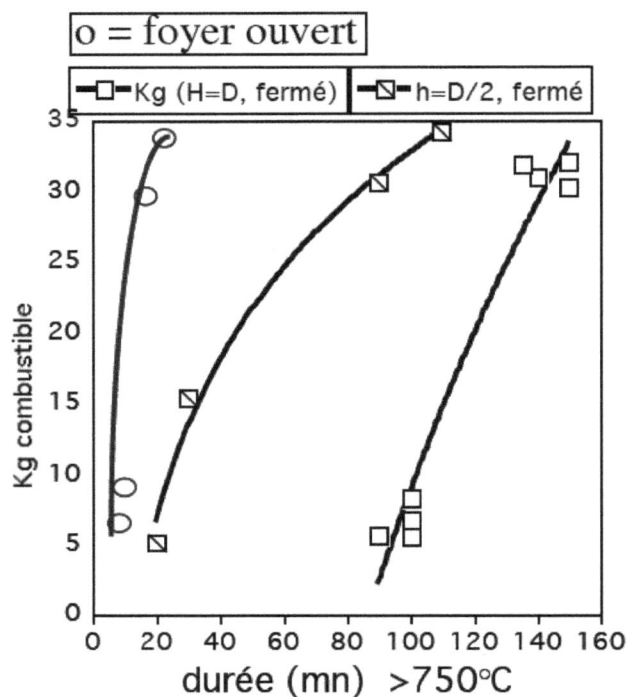

B

Structure		Chauffe	Tmax	t > 750°C	n
H=D	fermée	13±0.93°C/mn	919±51°C	121±25 mn	n=8
H=0.5D	fermée	13±0.82°C/mn	965±31°C	62±44 mn	n=4
O	ouverte	80 à 90°C/mn	950°C	5±5 mn	n=4

C

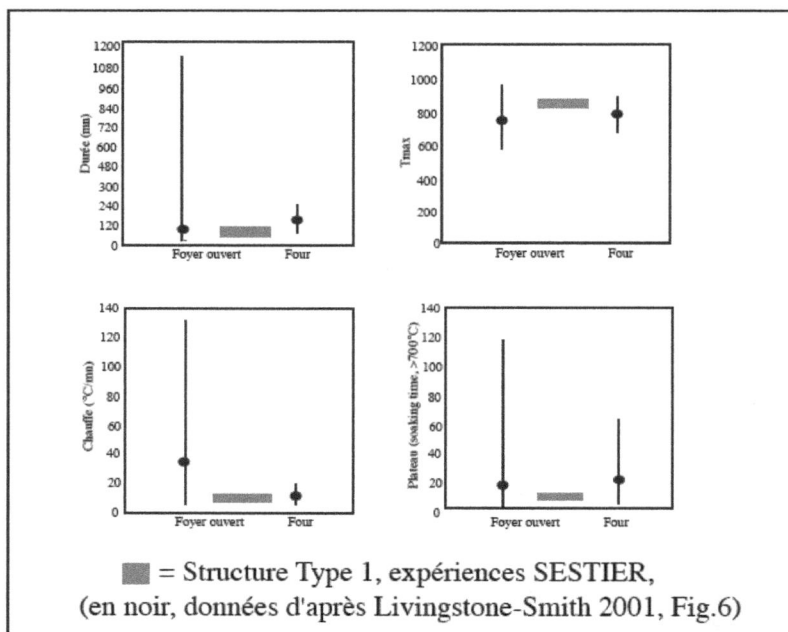

■ = Structure Type 1, expériences SESTIER, (en noir, données d'après Livingstone-Smith 2001, Fig.6)

Fig. 4: A) Interaction entre la structure de combustion, la durée du « *soaking time* » (>750°C) et la quantité de combustible. B) Résumé des résultats C) Comparaison du profil thermique de la structure 1 (en grisé) avec les résultats publiés par A. Livingstone-Smith (2001, fig. 6, p. 998), en noir, la valeur moyenne est figurée par un point entre les valeurs extrêmes (barre verticale).

Le savoir-faire a un rapport étroit avec sa transmission et la structure de la société, cette question est particulièrement intéressante lorsque diverses spécialisations apparaissent. Pour étudier de telle périodes de transition, il est nécessaire d'évaluer la faisabilité et la pertinence de l'analyse technique du matériel archéologique, pour formuler des hypothèses à caractère technique puis social (Pool 2000), car des procédés différents et concurrents peuvent coexister sans qu'on puisse les discriminer facilement. Enfin, la reconnaissance du «savoir-faire» facilite l'interprétation correcte des restes archéologiques souvent très fragmentaires et non-explicites, comme ceux laissés par des structures de combustion dites «primitives». Une partie de ce savoir-faire nous reste accessible de deux façons: par l'observation des pratiques actuelles (artisanales ou ethnologiques) et par l'expérimentation.

Remerciements: à M.-C. Frère-Sautot (APAB) et à l'Archéodrome de Beaune (Dijon, France).

Références bibliographiques

ARNAL, G-B.
  1989 *Céramique et céramologie du Néolithique de la France méditerranéenne.* Centre de recherche archéologique du Haut-Languedoc. Mémoire n° 5, Lodève, France.

ARNOLD, D.-E.
  1991 *Ceramic theory and cultural process.* Cambridge: Cambridge University Press.

BALKANSKY, A.-K., C. M. FAINMAN et L. M. NICHOLAS
  1997 Pottery Kilns of ancient Ejutla, Oaxaca, Mexico. *Journal of Field Archaeology,* 24: 139-160.

DREWS, G.
  1978 Entwicklung der Keramik-Brennöfen. Prague: *Acta Praehistorica et Archaeologica,* 9-10 (1978/9): 33-48.

DUHAMEL, P.
  1978 Morphologie et évolution des fours céramique en Europe Occidentale – protohistoire, monde celtique et Gaule romaine. Prague: *Acta Praehistorica et Archaeologica,* 9-10 (1978/9): 49-76.

GIBSON, A. et A. WOODS
  1997 *Prehistoric Pottery for the Archaeologist.* London: Leicester University Press.

GOUPY, J.
  1985 *La méthode des plans d'expérience. Optimisation du choix des essais et de l'interprétation des résultats.* Paris: Editions Dunod.

GOSSELAIN, O.-P.
  1992 Bonfire of the enquiries. Pottery temperatures in archaeology: what for? *Journal of Archaeological Sciences* 19(2): 243-59.

GOSSELAIN, O.-P. et A. LIVINGSTONE-SMITH
  1997 The «ceramic and society project»: an ethnographic and experimental approach to technological style. In: Lindhal, A. and O. O. Stilborg (eds.), *The aim of laboratory analysis in archaeology.* Stockholm: KVHAA Konferenser 34: 147-160.

KINGERY, W.-D.
  1997 The operational principles of ceramic kilns. In: Rice, P.M. (ed.), *Ceramic and civilisation Vol. VII. The prehistory and history of ceramic kilns.* Colombus: The American Ceramic Society, pp. 11-19.

LIVINGSTONE-SMITH, A.
  2001 Bonfire II: The return of pottery firing temperatures. *Journal of Archaeological Sciences* 28: 991-1003.

LÜDTKE, H. et R. VOSSEN
  1991 Töpfereiforschung-Archäologisch, Ethnologisch, Volkundlich. In: Hartwig Lüdtke und Rüdiger Vossen, *Beitrage des Internationalen Kolloquiums 1987 in Schelswig,* Band 2, Bonn: Rudolf Habelt GMBH.

ORTON, C., P. TYERS et A. VINCE
  1993 *Pottery in Archaeology,* Cambridge: Cambridge University Press.

PETRASCH, J.
  1986 Typologie und Function neolithischer Öfen in Mittel- und Südosteuropa. *Acta Praehistorica et Archaeologica* 18: 33-83.

POOL, C.-A.
  2000 Why a kiln? Firing technology in the sierra de Los Tuxtlas, Veracruz (Mexico). *Archaeometry* 42(1): 61-76.

RICE, P.-M.
  1987 *Pottery analysis. A sourcebook.* 2 vol, Chicago: The University of Chicago Press.

RYE, O.-S.
  1981 *Pottery technology.* Manuals of Archaeology, vol. 4, Washington: Taraxacum Inc.

SHEPARD, A.-O.
  1956 *Ceramics for the archaeologists,* Carnegie Institute of Washington.

VERTET, H.
  1978 Les fours de potier gallo-romains du centre de la Gaule. Prague, *Acta Praehistorica et Archaeologica,* 9-10 (1978/9): 145-157.

Preserved in Fire
Late Neolithic Settlement Structures in Western Hungary

Judit Regenye

Abstract

In Late Neolithic Transdanubia the connection of fire and clay is illustrated by three main instances, as ovens, pottery and fired wattle and daub houses. In all three, fire acts as a source of information to preserve the archaeological record by the conservation of cultural features, as for example the ruins of the fired houses, by conserving the imprints of rods together with those of boards, the angles of joining, the imprint of the posts and some other features.

Introduction

In the examination of the role of fire in Neolithic context, two approaches have to be differentiated: first the role fire played for the Late Neolithic people and second, the role of fire in the archaeological record by the conservation of finds and features in Late Neolithic archaeological sites. I chose the latter approach.

In course of the excavation of Neolithic settlements we often come across finds which survived to our day due to the conserving effect of fire.

The role of fire as a source of information is most apparent when it is reacting with clay, and the clay is becoming solidified by burning. In the remains of house walls we can come across definite prints of vegetal matter, traces of wooden construction material burnt out from the clay wall. On the basis of these we can draw technological conclusions about the structure of the houses. The ruins of the house remaining in situ can help in fixing the interior structure of the site, even without the traces of posts-holes.

Another element of settlement features closely connected with fire is the oven. The presence or absence of ovens on the excavated sites of the study area may have special significance. At the same time ovens may provide information on culinary practices as well.

Another notable case for the connection of fire and clay is pottery itself.

The sites to be discussed in the following paper were unearthed within the framework of the investigation of stone processing workshop centres in western Hungary. In relation to the study of mines and exploitation regions an intensive study of the workshop areas and the distribution zones was started in western Hungary (Transdanubia) in co-operation of the Laczkó Dezső Museum (Veszprém) and the Hungarian National

Fig. 1: The study area

Museum (Budapest) with the support of OTKA (the Hungarian National Research Fund) (Biró–Regenye 2003). Our interest was focused on the hill Tűzköveshegy (it means word for word "Flint Mountain") in Szentgál, because the site had a central role in the raw material supply of prehistoric Transdanubia. The Tűzköveshegy, a mountain in the Southern part of the Transdanubian Midmountains, is 438 meter high. The raw material exploited here is Middle Jurassic radiolarite (Biró 1995). Through examination of the stones found in the archaeological sites of the region, the relationship between certain Neolithic cultures and the mine at Tűzköveshegy became apparent (Biró-Regenye 1991).

Starting from the principle that mine is the part of the settlement structure, we did intensive surface investigations looking for Neolithic sites in a circle of 10 km radius around the hill. We found a settlement structure of 9 Lengyel culture sites around the mine; it was so unique as to make the researching of nearby sites of Lengyel culture necessary. Lengyel culture used to live, according to Hungarian chronological scheme, in the Late Neolithic and Early Copper Age (the Hungarian terminology according) ca. 4900-4000 BC between the Drava and Vistula rivers, the Alps and the Danube.

A series of excavations were planted around the mine. We excavated 4 of the 9 settlements (Szentgál–Füzi-kút, Szentgál–Teleki-dűlő, Ajka–Pál-major, Városlőd–Újmajor), all excavations were on a small area, our goal being to collect appreciable quantities of comparable material (Regenye 2001). The sites were undoubtedly in connection with the mine. Apart from the fact that the sites are geographically close to each other, the examination of the composition of stone materials excavated till now proved this relation. These sites can be considered as normal Neolithic villages where the chief source of living was the traditional Neolithic economy. In parallel, it is obvious that there was intensive connection to the radiolarite exploitation site and a definite specialisation on tool production, judging both from the actual location of the sites and the quantity and composition of the lithic industry (Biró 1993-94).

Preservation capacities of the fire

Some examples on the preservation capacities of fire from the excavation data of sites.

Case 1: The conservation role of destructive fire: remains of house walls.

In the excavations of prehistoric sites, more and more attention is paid to finds that were formerly considered less important, as plant remains, unprocessed stone finds and burnt daubing fragments of houses. They became important since they convey information on the life people lived in the settlement, on their past activities, completing the data gained from the primary find assemblage of a site, the household garbage (pottery fragments, animal bones, stone tools).

The burnt daub with his high informative value deserves attention in particular among the above mentioned 'less important' find types. Daubing fragments are frequent to be found in Neolithic settlements: the reason is that the timber structure of the houses is inflammable. The open fireplaces, the splitting sparks can be the cause of conflagration in a settlement, and the occasional intentional burning down of the houses is also supposed. The explosive fire burns out the clay plastering of the walls and so preserves signs, indicating the wooden elements of the house structure. Technolecal observations can be made regarding to the building structure. Fire preserves not only the finds themselves, but even fixes their spatial position.

The house type where the wattle-and daub walls rest on a timber structure was invented and spread in Central Europe by the Linear Pottery Culture, which also introduced cultivation in the 6th millennium BC. The applicability of the building material was determined, beside lifestyle, by the natural environment, the soil conditions and climate, especially the amount of the precipitation. Houses with wattle-and-daub constructions fitted excellently the Central European climate. This is why it was used throughout the prehistoric times, as it can be deduced from the archaeological finds. It reappeared in the Middle Ages and the construction technology was observed in folk architecture as well. The wooden structure of Neolithic houses can't be ever studied by direct means but it can be illustrated by examples taken from the folk architecture. These illustrations can help us in data recording of imprints on burnt daubing fragments. Pictures from the thirties show the wattle as building element, even a church has wattle-and-daub wall (**Figs 2-4**).

Fig. 2: Country house, built with wattle-and daub walls (Kolontár, Hungary, 1937).

It should be added that daub was used not only in wattle-and daub constructions. A wide range of earth-and-wood constructions is known in the Hungarian folk architecture and the picture was probably equally colourful in prehistoric times as well.

Saddle-roofed long houses, which were usually made with daubed walls, retained dominance in folk architecture until recent times. The long houses of the Linear Pottery culture, regarding their proportions, were very similar to the peasant houses of recent times. In later phases of Prehistory and even in the early medieval period, the dwellings were much smaller. The monumentality of the Neolithic houses (first of all of the Linear Pottery Culture houses) is conspicuous in

Fig. 3: Detail of a barn (Csetény, Hungary, 1937).

Fig. 4: Church with wattle-and daub-built walls (Kemse, Hungary, 1925).

The clayey soil that covers the examined territory in western Hungary makes the recognition of the subterranean elements of the house remains difficult at the excavations of Neolithic settlements. Often only the demolition layer, the burnt remains of houses indicate the presence of a dwelling.

Fig. 5: Demolition layer of a Neolithic house (Kup, Hungary).

Depending on the intensity of fire that devastated the house, the wall remains can be slightly or thoroughly burnt. Usually, the clay daub does not burn through so erosion disintegrates the fragments leaving only a reddish discoloration in the soil.

The strongly burnt items, however, resist erosion and preserve the imprints of the wooden elements that compose the skeleton of the house. The walling technology can, to a certain extent, be reconstructed from these imprints, and the reconstructed details can be compared with the methods applied in folk architecture. In the unearthed Late Neolithic settlements, most of the items show the imprints of branches, which attests to a wattle made of rods as thick as a finger (**Figs 6-8**).

The imprints of split boards or laths appearing together with the imprints of wattle suggest that boards were used for the reinforcement of the construction (**Figs 9-11**).

relation to these latter ones, and the huge buildings must have had a symbolic value beside the use-value. Houses later then the Neolithic were usually smaller, the long house is an extreme case, they matched the everyday needs better. The symbolic role, if it ever existed, was pushed back, perhaps it had lost its meaning. This difference was not manifested, however, in the construction technology, all the houses were built with the same technology irrespective of their measurements. It was true at least for the wall construction.

Fig. 6: Daub fragment with imprint of rods.

Fig. 7: Daub fragment – reconstructed situation.

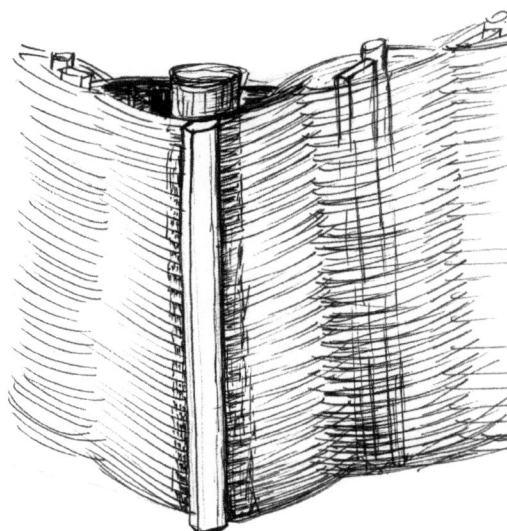

Fig. 10: Reconstructed wattle structure of the wall.

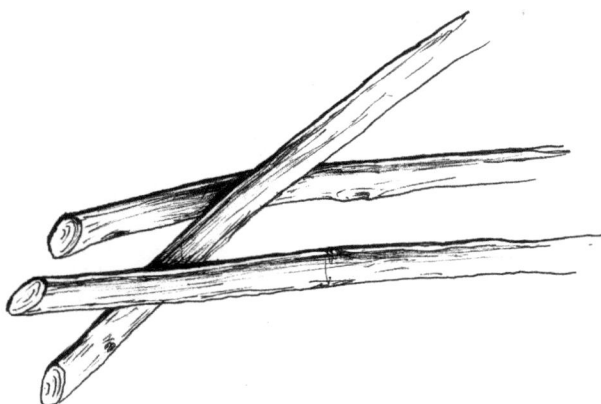

Fig. 8: Reconstructed wooden element of the house wall structure - wattle made of rods.

Fig. 11: Reconstructed wooden elements of the house wall structure.

Astonishingly many board imprints can be observed in the daub remains found in Szentgál–Teleki-dűlő (**Fig. 12**), one of the Lengyel culture settlements around the stone exploitation place mentioned above. We excavated a house with a demolition layer rich in daubing fragments in good condition.

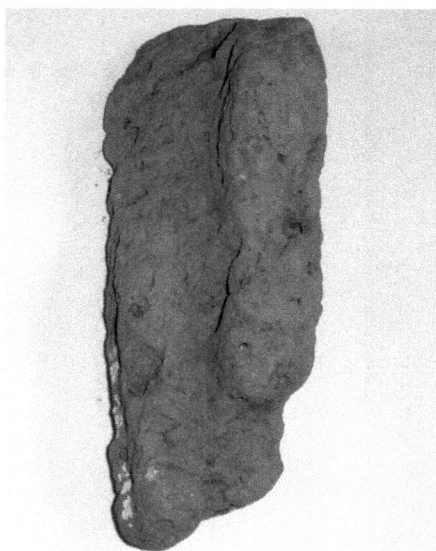

Fig. 9: Daub fragment, imprint of rod and board.

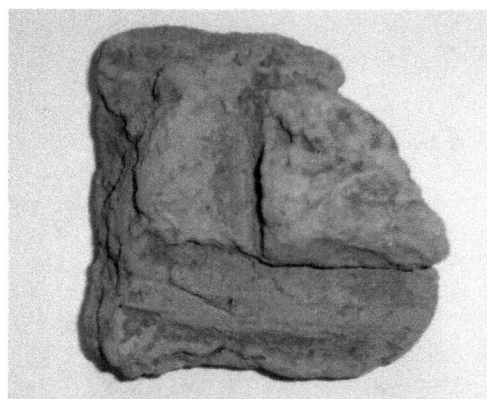

Fig. 12: Daub fragment, imprint of board.

Nearly every well preserved daub fragment showed the imprints of rods together with those of boards, often at a right angle to one another. In these cases the boards were probably placed into the gaps of the wattle for reinforcement. The items that preserved the board imprint on the outside can imply that one side (probably the inner one) of the wall was covered with boards (**Fig. 13**).

Fig. 13: Daub fragment, imprint of board.

The board imprint running obliquely beside a post imprint on items where rod imprints are missing implies that not only wattle could provide the skeleton of a wall. Other technologies known from finds and descriptions from the 13th-14th centuries could also be applied as e.g. daubed post structure or rammed cob walls.

Fig. 14: Daub fragment, imprint of piles.

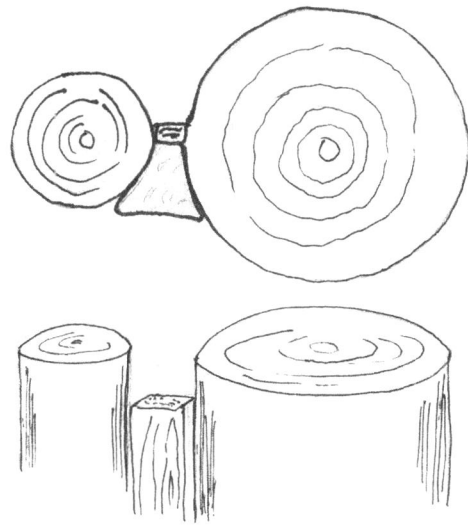

Fig. 15: Reconstructed wooden structure on the basis of the previous daub fragment.

Fig. 16: Reconstructed wooden element of the house wall structure – daub between boards.

Fig. 17: Daub fragment – reconstructed situation.

A cob wall can be reconstructed from a few fragments (**Figs 16-17**). In this case earth was rammed between boards supported by posts. When the wall segment dried out, the boards were raised higher, leaving the board imprints in the dried earth. Sometimes sticks were also rammed into the wall, in this case the imprint shows the rod and board imprint described at the wattle-and-daub wall. In a few cases the earth was rammed between two wattles, then the wattles were daubed from both sides. The remains of these variants are difficult to identify in the archaeological record, since their small fragments appear to be very similar. Some daub fragments are regularly arched on one side. In rare cases the diameter of the prop can be calculated from the intact arch of these daub fragments (**Figs 18-21**).

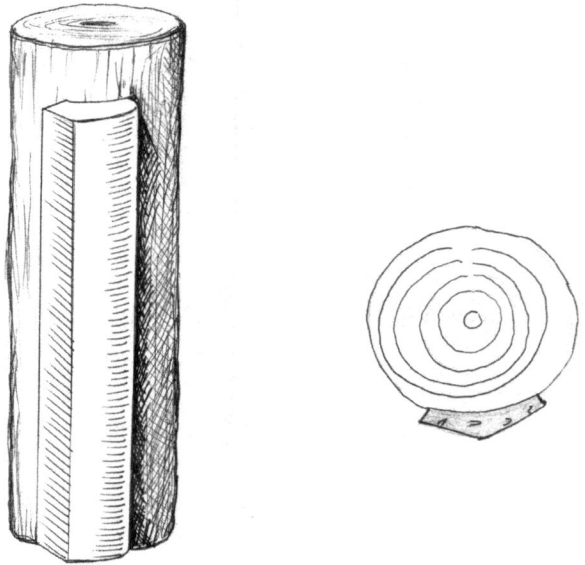

Fig. 21: Reconstructed position of the previous fragment.

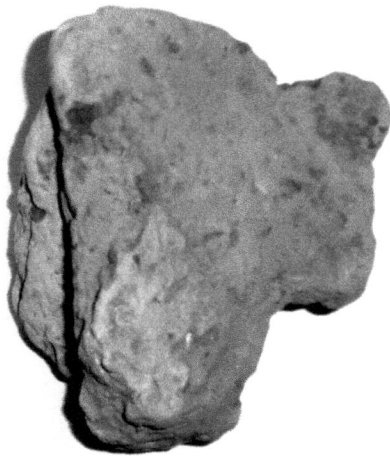

Fig. 18: Daub fragment, imprint of a post.

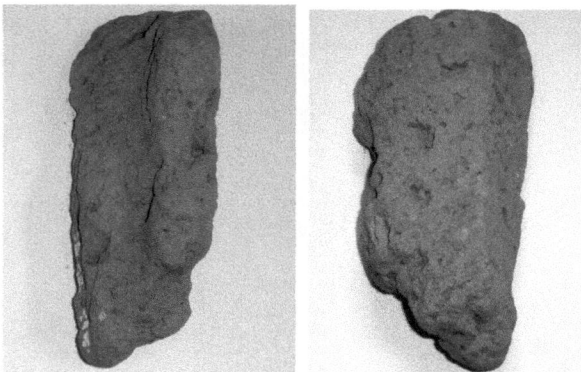

Fig. 19: Daub fragment, imprint of a post.
Fig. 20: Daub fragment, the back side of the previous item.

Hungarian peasant houses in the Hungarian Plain evidence that the load-bearing capacity of mud walls is great, they are able to hold the roof without props. When postholes of small diameters are found along the side walls we can presume after this example that the smaller posts did not support the roof, only reinforced the walls.

The measurements of the fragments do not reflect the entire width of the wall since erosion wears off the surfaces. A reconstruction experiment implies that the thickness of the wall could reach 20-30 cm. The wattle itself was about 10 cm thick depending on the thickness of the reinforcing elements. The use of thicker reinforcing elements was not advantageous since the limited elasticity of the twigs would only have afforded a loose structure, and a loose wattle is difficult to daub. The entire width of the wall was reached by the daub applied on the wattle from both sides. As far as I could observe, the number of the daub fragments found in a demolition layer is only a portion of the mass of clay necessary for the daubing of a house. The discrepancy is due to the weathering of the daub and the damage done by the plough.

The presence of daub at a site can, in itself, play a crucial role in the determination of the type of the site. A Lengyel culture settlement was discovered during field walking at Zirc within the frames of the investigation of stone processing workshops. The site lay in a valley of the Bakony hills 400 m above sea level, somewhat separated from the above mentioned group. Stone finds dominated among the surface finds, namely stone blocks from a nearby raw material provenance.

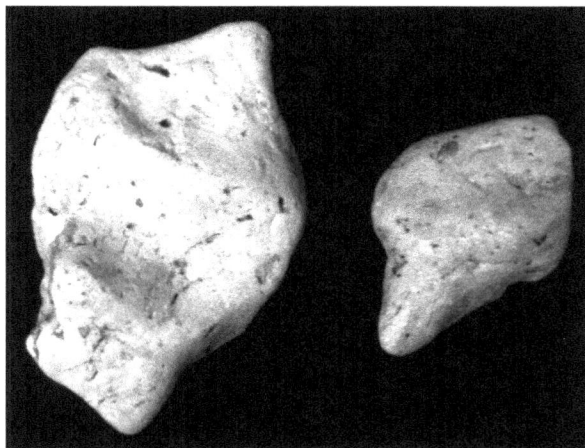

Fig. 22: Daub fragment, surface finds.

The pottery material we collected on the surface was small in number, it was an average settlement material. Based on the daub fragments (**Fig. 22**), there stood a permanent house in the settlement, which means that it was not a temporary shepherds' or miners' camp as it could have been supposed at an elevation where usually no Neolithic settlements occur. The fact that the people of the Lengyel culture so strongly clung to the raw material provenances resulted, elsewhere as well, that they founded settlements near the raw material sources even within extreme circumstances and in areas that were generally neglected during the Neolithic. In this case the daub fragments were the best arguments for the existence of a permanent settlement in the middle of the Bakony Mt. and could help in recording a characteristic feature of the Lengyel culture settlement system.

Case 2 of the interaction of earth and fire: Ovens

Ovens were important elements of everyday life (heating, cooking, baking, desiccating), and, at the same time, they were architectural constructions. They are mentioned here because of their absence. It means that we can rarely if ever discover any trace of an oven or a hearth on excavations of the Late Neolithic in the examined territory. This is not a unique experience: only 27 of the 160 analysed houses published in a summary on Central European Prehistoric house construction in 1992 contained information on hearths (Luley 1992: 8). Survival is impossible within the Central European climatic conditions without some form of heating, and the existence a fire place must be presumed because of other functions (cooking etc.) as well. The lack of ovens in the sites of the examined territory can perhaps be explained by the small surface the excavations covered, although it included an entire house as well. We can also suppose that due to some reason the ovens became archaeologically unobservable (e.g. they were built higher than the floor lever or the floor level has perished etc.). This latter is justified by the fact that fragments of oven plastering were observed in pits in several cases. In a recently unearthed

settlement of the Transdanubian Neolithic (Kup), an oven dug into the earth yielded interesting data. Fist-sized pebbles, which do not occur naturally at the site, were found arranged in a row at the mouth of the oven (**Fig. 23**).

Fig. 23: Oven with stones.

The logical supposition that first emerges is that they functioned as cooking stones, although the pebbles placed at hand next to the oven could be used for other purposes as well. It is well-known that liquid can be boiled with the help of stones, which was probably applied in the Neolithic as well, even though people could use pottery vessels for cooking.

Case 3: Ceramics

Another group that bears information preserved by fire is ceramics. Regarding the role of fire in conservation, spontaneous and intentional preservation can be differentiated. As opposed to houses that got unintentionally burnt down, clay was made resistant by the conscious use of fire.

The potsherds are the most frequent finds in a site; no other finds are as informative, except probably the lithics. The characteristic of pottery is the continuous changes through time that is the reason why we can identify archaeological periods on the basis of ceramics. Another outstanding characteristic of pottery is its near indestructibility. In form of potsherds they are no more easily broken, and once broken, the potsherds are likely to be preserved, since they can be destroyed only by severe erosion.

There is another difference between the position of daubing fragments and the ceramics. Unlike demolition debris that stayed in its original place until the archaeological excavations, the pottery fragments were discarded and considered garbage, they are found elsewhere than the place where they were used. The demolition layer of houses sometimes preserved the household furnishings when the house was destroyed by a quick fire, yet most of the pottery remains are discovered where they were discarded in fragments. It is a common phenomenon that the fragments of the same vessel are found at various parts of a Neolithic settlement, which means that the garbage was not put directly into the pits but it was moved before arriving there. Starting from modern rural practice, the most evident supposition is that the broken vessels were put in the dump, which was later pushed into the pit. It is, anyhow, certain that a habitual behaviour can be discovered in the discarding of things, which is reflected in the distribution of the refuse in a settlement.

The investigation of human behaviour is based, from this respect, on the garbage and fire with its conserving properties contributed to the preservation of information we need for this investigation.

Bibliography

BIRÓ, K.
  1993-94   Szentgál-Füzikút késő neolit település kőanyaga (Lithic material of the Late Neolithic settlement Szentgál, Füzikút.) *Veszprém Megyei Múzeumok Közleményei* 19-20: 89-118.
  1995   H 8 Szentgál-Tűzköveshegy, Veszprém County. *Archaeologia Polona* 33: 402-408.

BIRÓ, K. and J. REGENYE
  1991   Prehistoric workshop and exploitation site Szentgál-Tűzköveshegy. *Acta Archaeologica Hung.* 43: 337-375.
  2003   Exploitation regions and workshop complexes in the Bakony Mountains, Hungary. In: Stöllner, Th., G. Körlin, G. Steffens and J. Cierny (Eds.) *Man and mining – Mensch und Bergbau. Studies in honour of Gerd Weisgerber on occasion of his 65th birthday.* Bochum, pp. 55-63.

LULEY, H.
  1992   Urgeschichtlicher Hausbau in Mitteleuropa. Grundlagenforschung, Umweltbedingungen und bautechnische Rekonstruktionen. *Universitätsforschungen zur Prähistorischen Archäologie*, Band 7. Bonn.

REGENYE, J.
  2001   Settlements of the Lengyel cuture around the Tűzköveshegy in Szentgál. In: Regenye J. (ed.) *Veszprém Sites and stones. Lengyel culture in western Hungary and beyond*, pp. 71-79

Chalcolithic Pyroinstruments with Air-Draught – An Outline

Dragos Gheorghiu

Abstract

The present text discusses the analogy of the process of air-draughts in different Chalcolithic ceramic pyro-objects with perforated surfaces. Such special designs for increasing the air draught allows the identification of pyroinstruments in archaeological records because of their openings and perforations, as individual objects or as interrelated assemblages.

Introduction

Fire is at the same time a phenomenon and material culture. Usually, in archaeology, fire is studied under its materialized form, by analyzing its material supports which confer to the phenomenon its instrumentality (that transforms it into a material tool, or pyroinstrument).

Of all investigative methods, experimentation[1] allows for a more nuanced understanding of the phenomenon, revealing a multitude of its aspects otherwise impossible to perceive solely from a theoretical approach. Physical principles or rituality, to cite some examples, are cases in point, since traditional people interacted with fire in a highly ritual manner (Budis 1998: 66 ff). Thus experimentation with pyroinstruments is the most appropriate instrument to study ancient fires. An additional aspect that may be studied through experimentation is the holistic relationship of fire with the household.

In spite of the individuality of pyroinstruments found in the prehistoric household, the use of fire was a holistic action which developed from micro to macro scale.

The present paper intends to present the relationships of fire with clay objects within the southeastern European Chalcolithic traditions by analyzing the pyro-objects that function using air-draughts, from small utilitarian clay objects up to the house's inner space.

Straining, sieving and Chalcolithic pyrotechnology

I believe the development of Chalcolithic pyrotechnology is the result of the "second products revolution" (Sherratt 1981), since there seem to be technological analogies between food and firing processes. Such an interpretation is based on the presence in the archaeological record of several types of ceramic objects with perforations, some

of them used in the dairying process and some with the role of pyroinstruments.

In a Chalcolithic house, from micro to macro objects, the control of fire used the same principle of air-draughts. All air-draught pyroinstruments shared the property of suctioning air through their bodies. Such a process could be initiated if the surface of the objects was perforated so the air could circulate and feed the fire positioned inside. A basic definition of the phenomenon of air-draught would then be the property of a heated pyroinstrument to create an ascending flow of hot air that would carry the flames of fire.

Ethnological models infer that in a household the functions achieved by pyroinstruments could have been the following:

- starting the fire and preserving it as embers during the night;
- preparation, drying, heating, smoking, desiccation and conservation of food;
- heating the interior of the house;
- illumination;
- evacuation of the smoke produced by all pyroinstruments within the household.

Outside the household mention should be made of the pyroinstruments' role in transforming inorganic materials.

To identify the above-mentioned functions, experimental archaeology seems to be the most appropriate instrument to illustrate the dichotomy between shape and function found in the archaeological record. As a result of experiments the above-mentioned functions were identified as being represented by the following objects:

- fire starters and preservers of embers in the shape of ceramic cones with perforated walls;
- ovens with vents and ceramic shutters;

---

[1] The present paper describes experiments conducted through the Vadastra Project, between 2000 and 2004, within the Experimental Pyrotechnology Group.

- liquid heaters, seed or salty solution desiccators in the shape of ceramic vases with perforated supports;
- food smoking container in the shape of an attic with a ceiling aperture;
- braziers in the shape of perforated cylinders or rectangular prisms, sometimes designed to symbolize architectural objects;
- macro pyro-objects formed by the functioning of all the pyro-objects of houses;
- the macro pyro-object produced during the process of intentional firing of the house, and
- up-draught kilns with perforated platform (positioned outside the household).

Fire starters

The action of processing vegetal or milk products through sieving or straining are inferred in the archaeological remains by the ceramic vases with perforated bottoms, often found in the Lower Danube region dating from the 5th millennium B.C. Until recently archaeologists assigned all Chalcolithic and EBA objects with perforated walls to the category of food strainers, especially the objects with perforated corbelled or conical walls. The objects are of varying dimensions, from small (Cucuteni tradition, Tarpesti, h = 4 cm; Museum Piatra Neamt), to large (Poduri, h = 38 cm, Piatra Neamt Museum), and have a quite large perforation on the narrow side. Jane Wood identified these corbelled sieves as being prehistoric "Bunsen lamps" and experiments (see Wood 1999, Gheorghiu 2002a) demonstrated they are capable of producing a flame when put on embers. This new perspective on Chalcolithic design would probably reclassify many objects with perforated walls (see Gheorghiu 2002a: 88; Gheorghiu 2003: 41, fig. 5).

Following repeated experiments my interpretation of these objects' role is that they were used as "ember protectors" or "fire starters" / "fire regenerators", because they can restart a fire through the simple act of blowing air on it (**Fig. 1**).

A second function of these objects would have been to act as lamps when the air blown on the perforated walls produced a strong flame for a short interval of time.

A third function could have been that of "smokers", used less to preserve aliments than to protect the inhabitants from insects, the smoke of the embers coming through the perforated surfaces in the absence of air-draughts.

Ovens

Ovens were combustion structures (see Gheorghiu 2002a: 87, fig. 6), sometimes with symbolic shapes (see Comsa 1990) that functioned as air-draught instruments due to the perforations positioned at the upper part, opposite to

Fig. 1: Replica of a fire starter. Scale 20/1. Vadastra campaign 2002. (Made by Cristian Mare)

the main opening, as shown by miniature models (Gheorghiu 2002a: 87, fig. 7).

Experiments showed that a good control of the air-draught through an oven can be achieved by partially covering the upper orifice(s).

Ovens' shutters

While trying to make bread in an oven with a vent[2] and trying to control the firing process, I discovered the importance of shutters in achieving a good air draught in a large pyroinstrument. To be functional, a shutter must have been made of clay and have been very ergonomic. The most obvious examples found in the archaeological record are the horse-shoe or round ceramic grates with partial perforations (Dumitrescu *et al.* 1954; Ellis 1986: 374), quite frequent in the Cucuteni tradition, identified first as being "kiln grates" (Dumitrescu *et al.* 1954: 189-198). Their small dimensions (e.g. 48.5 x 34 x 4 cm at Habasesti) correspond more to the dimensions of the mouth of an oven and not to the dimensions of the firing chamber of an efficient kiln. The incomplete perforations had the functional role of positioning the heated shutter over the mouth of the oven with the help of fingers or of wooden sticks (**Fig. 2**).

---

[2] The model was built by Alexander Chodaziev with the help of villager Constantin Liceanu in 2003 in Vadastra.

Fig. 2: Oven shutter. Vadastra campaign 2003.

Fig. 3: Vase with a perforated support. Cucuteni tradition. Piatra Neamt Museum.

the geometric-anthropomorphic supports (Cucuteni tradition, Poduri tell, Piatra Neamt Museum).

Fig. 4: Vase with perforated bottom and four small arms. Gumelnita tradition, Sultana tell, Calarasi Museum.

Fig. 5: Replica of the Sultana vase. Vadastra campaign 2003. (Made by Andreea Oprita)

Liquid heaters, seed or salt solution desiccators

Perforations could also be found on plates or pot stands, in this case their function being to allow the hot air to circulate under the vase positioned on embers, and heat the liquids, semi-liquid food, or salty solutions, or desiccate seeds or food. This type of vase emerges in the ceramic assemblages of Early and Middle Neolithic South Eastern Europe, such as in the Franchti Cave (see Vitelli 1995: 57, fig. 5.2).

In Eastern Europe this type is to be found as early as Starcevo-Cris (Carcea and Gura Baciului) (Lazarovici and Maxim 1995), becoming frequent in the Gumelnita and Cucuteni-Tripolye traditions (**Fig. 3**), where numerous variants are to be found, such as the cone-shaped type with four arms (Gumelnita tradition, Sultana tell, Calarasi (**Figs 4 and 5**) and Oltenita museums), or

Both types of objects cited were formed from two separate parts, a stand with perforation(s) and a vase, a design that developed in the Cucuteni tradition to a single cylindrical compact shape with the vase slightly profiled. One could find in a single Cucuteni A-B (middle phase) tell settlement like Izvoare several dimensional types of the same compact object with varying volumes, from small (diameter 15 cm, h. 40 cm) to middle (diameter 25 cm, h. 55 cm), to large (diameter 40 cm, h. 55 cm), all displaying standard perforations of 2 cm in diameter at the upper part of the stand.

Fig. 6: Replica of the ceramic object discovered at Cascioarele tell, Gumelnita tradition. National Museum of History, Bucharest. Vadastra campaign 2003, (Made by Andreea Oprita)

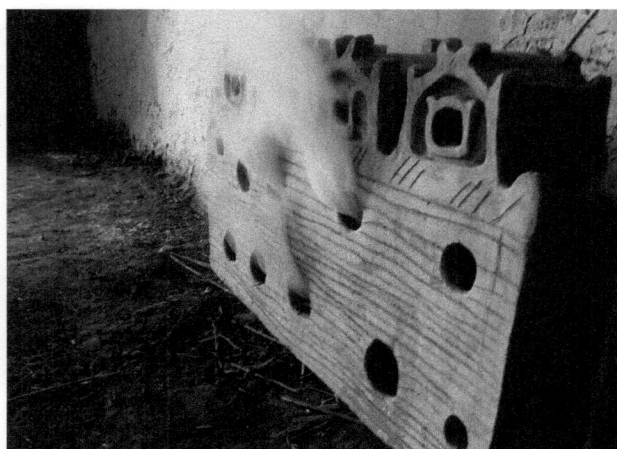

Fig. 7: The ceramic object from Cascioarele used as a brazier. Vadastra campaign 2004.

All the vases with perforations on their stands were modelled from a fine paste and coated with a fine grained slip, or were painted, some of them also on the interior of the upper container, which infers that the substance heated would have had a special ritual value, and also that the vase was not frequently used. One possible function of some of these objects in the Starcevo-Cris (Carcea and Gura Baciului) (see Maxim 1999: 31) and Cucuteni traditions would have been the heating of opiate liquids, since the shape of the upper container looks analogous to a poppy capsule.

Braziers

Particularly in the Gumelnita tradition there is a large corpus of unidentified ceramic objects with perforations whose function is not yet identified, that are generally

classified as "cult objects". A famous one is the "altar" or "temple" model discovered on the Cascioarele island (L = 0.5 m, l = 0.13 m, h = 0.24 m; Dumitrescu 1968), illustrating a row of megaron houses positioned on a stylobate with perforations (**Fig. 6**). In my opinion this symbolic object functioned as a brazier and illustrates the image of a tell site, surrounded by a palisade, the smoke going out with the air draught offering an authentic impression of an inhabited settlement (Gheorghiu 2003: 41, fig. 5; Gheorghiu 2002b: 103-104).

In the summer of 2004 I carried out experiments with a replica of this model.[3] When the fire was lit inside, the ceramic box evoked the image of a settlement with rising smoke from many households (**Fig. 7**).

Other analogous objects, in the form of a dense settlement positioned on a rectangular box with perforations (L = 0.25 m, l = 0.08 m, h = 0.28 m; Serbanescu 1997: 249, fig. 3/5; Gheorghiu 2002b: 104), or in the form of a house with perforated walls and roof (L = 0.32 m, l = 0.17 m, h = 0.21 m; Serbanescu 1997: 235, fig. 2/4; 4/2) were discovered in the eponymous Gumelnita settlement and in Sultana (Oltenita Museum) (**Fig. 8**).

Fig. 8: House model from Sultana tell, Gumelnita tradition. Oltenita Museum.

To the same category of symbolic braziers from the Gumelnita tradition belongs a ceramic architectural model discovered in Drama (see Lichardus *et al.* 1996; Gheorghiu 2002a: 87, fig. 7) which has large perforations situated on its back side, probably because this brazier copies the shape of an oven, with the openings positioned opposite the mouth opening to generate an efficient air-draught. A similar category, including room heaters and vase-heaters, also comes from the Gumelnita tradition, materialized as ceramic cylinders whose upper surface has a number of round openings serving for air draught and to support vases (e.g. from Ulmeni settlement, one object with a diameter of 45 cm, h. 40 cm, with 5 openings, one object with a diameter of 25 cm, h. 30 cm, with one opening a diameter

---

[3] Model designed by student Andreea Oprita, National University of Arts in Bucharest.

of 18 cm, one object with a diameter of 30 cm, h.18 cm, with one opening with a diameter of 22 cm, all decorated with spirals, Oltenita Museum).

Up-draught kilns

I believe that up-draught kilns are the most eloquent example for demonstrating the instrumentality of fire through a perforated material support.

Draught kilns are attested in the archaeological record from the 6th millennium B.C. in the Near East (see Simson 1997: 39), but only in the 5th - 4th millennia B.C. in South Eastern Europe (see Markevici 1981 and Comsa 1976: 25).

Their invention occurred probably very early, likely before the emergence of ceramics in the Near East, and could have been the source of the *vaisselle blanche /* white ware (Contenson and Courtois 1979), produced by coating wickerwork containers with lime plaster (Gourdin and Kingery 1975), the latter material requiring high temperatures which could have been achieved in draught kilns.

The emergence of draught kilns corresponds to an increased mastery of fire through the control (see Rye and Evans 1976: 164; Arnold 1997: 213) of the ascending hot gases flowing through perforated surfaces with an air-absorbing character. This, in turn, was the result of an increase in ceramic production and the emergence of craft specialization. To understand the functioning of up-draught kilns, the perforated platform can be viewed as a set of air absorption tubes, whose role was the distribution of the thermal shock (**Fig. 9**).

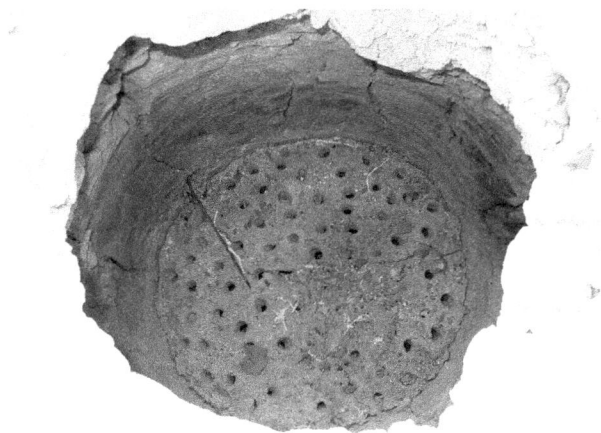

Fig. 9: The perforated platform of an up-draught kiln. Replica of a Cucuteni phase B kiln. Vadastra campaign 2002. (Made by Dragos Gheorghiu and Alex Gibson)

The key operations of the draught kiln are as follows: gradual evacuation of capillary water (Arnold 1997: 61),

heat containment (Shepard 1956: 75; Kingery 1997: 11) and heat transfer to wares, protection of ceramic objects contained by the firing chamber from thermal shocks during the processes of heating or cooling, protection against humidity (rain or moisture) (Arnold 1997: 213) and the production of higher temperatures (Shepard 1956: 75) as compared to other pyroinstruments.

Depending on the direction of the air flow there are down-draught and up-draught kilns, the latter being mentioned in the archaeological record of the late Eastern European Chalcolithic. The following parts form the functional shape of an up-draught kiln: the fire tunnel/firebox, the firing chamber, a perforated platform, a support for the platform and ceramic waste to cover the vent.

The differences between the open structures of combustion and the up-draught kilns consist of the following features:

- the fuel is separated from the clay objects which are not in direct contact with the flame;
- the inner space of the pyro-object is divided by a perforated platform into a fire box and a firing chamber;
- the thermal shock of the hot gases and flames is strained and sieved by the tubes of the perforated platform;
- a regulation of the amount of fuel in the firing box occurs in order to control the interior temperature;
- an opening or a closing of the apertures of the kiln (Rye and Evans 1976: 164) having as final the result a control of the oxygen, an operation which can create an oxidizing or a reducing atmosphere.

All these features characterized by the notion of "control" (control of air flow and consequently control of fire by changing the time and temperature of firing by means of manipulating the fuel, or control of the atmosphere by closing the apertures), convert the up-draught kiln into a "machine", i.e., a complex instrument with inputs and outputs, which can be regulated.

Chaîne-opératoire. Determinisms and subjective choices

The replicas of the up-draught kilns used for experimentation were copied after the kilns discovered at Costesti (Markevici 1981) and Glavanestii Vechi (Comsa 1976: 25) dated from the final phase of the Cucuteni-Tripolye tradition (5th – 4th millennia B.C.).

Their shapes can be reduced to the following basic volumes: a semi-cylindrical fire box (1.5 m x 0.5 m), intersected with an ovoid (diameter at the base of 1.5 m, depth 2 m), divided by a clay platform (height: 0.45 -

0.50 m) with perforations (diameter: 3 - 5 cm). The ovoid shape seems to have been the most efficient in creating a good air-draught and conserving thermal energy during and after the combustion process; the difficulty it presents pertains to the loading and unloading of the ceramic objects, an operation probably carried out with the help of children (**Fig. 10**).

A first set of operations was the loading of the firing chamber through the vent (Fig 11; Fig. 12), followed by the covering of the vent with shards, to form a ceramic strainer (Fig. 13), and the structured positioning of the fuel in front of the fire box (Fig. 14). Both fuel and covering shard were positioned to allow the creation of an airflow between the fire box and the vent.

capillary water (Arnold 1997: 61) (Fig. 16). There is a relationship between these stages and the dimension of the kiln and of the pieces which form the load.

Fire temperatures can be raised to over 350°C, the limit of combustion for the organic material contained in the clay's fabric, but stabilized under 700°C, which represents the second stage of firing, when the process of sintering of the internal particles began (Kingery 1997: 12). At temperatures between 570°C and 600°C the elimination of water from the crystal structure of the minerals began (Shepard 1956: 72; 81; Arnold 1997: 610), which is associated with the start of red light emissions from the incandescent clay objects. The heating at over 700°C of the interior of the kilns and of the ceramic load produced a strong air draught, which allowed the flames from the fire box to reach up to the vent.

Fig. 10 : Loading the kiln with the help of children. Vadastra campaign 2002.

Fig. 12: The kiln loaded with vases. Vadastra campaign 2003.

Fig. 11: Loading the kiln through the vent. Vadastra campaign 2003.

After initiating the fire, the fired fuel was gradually introduced inside the fire box (**Fig. 15**), the fire being stabilized between 20°C and 120°C (see Hamer 1975: 22), to allow water smoking, i.e. the slow elimination of

Fig. 13: The cover of the vent. Vadastra campaign 2003.

Frequently, when the temperature reached 900°C at the middle of the load, the fire box was already filled with embers and the air draught diminished, consequently impeding a further rise in temperature.

One can infer that the volumetric structure and the dimensions of this type of kiln are optimal for temperatures under 900°C. To elevate the temperature over this thermal threshold one should discharge the firebox and initiate a new fire which would help to go over 1000°C.

Fig. 14: The structured positioning of the fuel in front of the fire box. Vadastra campaign 2002.

Fig. 15: Firing the kiln. Vadastra campaign 2003.

Since this is a procedure requiring high energy expenditure and comprising a high degree of technical difficulty, it is possible that high temperatures in draught kilns were not reached in the East European prehistory, as supported by the chemical analyses of prehistoric ceramics.

At temperatures over 1150°C (when using refractory clay for building the platform) one notices the beginning of a process of vitrification in the fill, with glass and slag in the air-draught tubes/perforations, as a consequence of the rise in the temperature.

Beside the stages of water elimination, which require the observance of a well determined time, the other stages of the firing process could be done using different spans of time, therefore a complete process of firing could last between 5 and 18 hours (see Deca 1982: 214), depending on the type of load, fuel and temperature of the environment, but also on the style of firing of the potter.

Fig. 16: Water smoking. Vadastra campaign 2002.

The house as pyroinstrument: Food smoking

A draught effect similar to that of ovens could be created in large architectural spaces. Experiments carried out during the summer of 2003 in the Vadastra archaeopark with a full scale architectural object,[4] revealed that the household space could act as a large pyroinstrument, in aspirating and evacuating smoke through the opening of the window and the ceiling, and that a second role of the attic, after that of temporal residence, would have been that of food smoking during the cold season, since the thatched roof's permeability would produce an air draught which would absorb the oven's and other

---

[4] The replica of a Chalcolithic house was built in summer 2003 by the author with the help of art students (National University of Arts in Bucharest) and Vadastra villagers.

pyroinstruments' smoke through the ceiling opening and evacuate it through the roof's texture (**Fig. 17**).

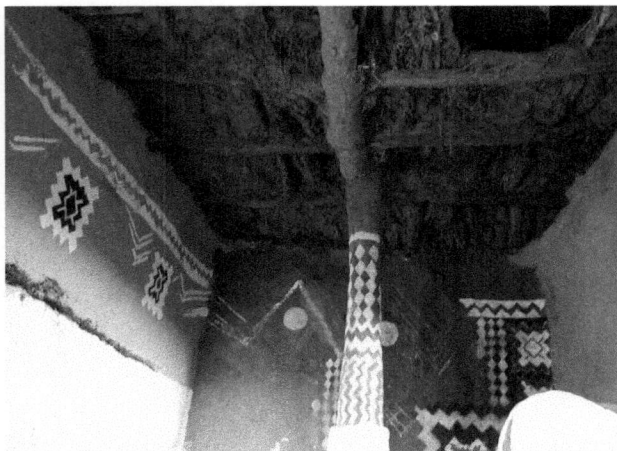

Fig. 17: Replica of a Chalcolithic house from Radovanu tell, end of Boian tradition. Vadastra campaign 2003. One can see the ceiling opening on the right side of the image.

Control of the air draught could have been achieved by partially or totally closing the mentioned apertures with shutters similar to those of ovens.

In the relationship between the household and fire two main seasonal episodes can be identified: during summer fire was external to the house, and materialized as ovens and various pyrostructures, built around the house (Marinescu-Bilcu *et al.* 1997), while in winter fire was positioned in the core of the inside activity area, under the shape of the oven, and also spread all over the inner space, contained by small pyroinstruments like braziers, food heaters and other small objects.

Since all the pyro-objects discussed were designed to sieve the air, one could perceive the household-*oikos* as a place of sieving and modelling both flour and fire.

The house as pyroinstrument: Intentional firing

Studies of South Eastern European Neolithic settlements have tried to interpret the frequently fired horizons as the result of a performance ritual of intentional firing of houses (Tringham 1992; Stefanovic 2002). Although the subject seems at first glance unrelated to the theme of the present paper, the two share a common point as far as the use of air draughts is concerned. As wattle-and-daub Chalcolithic houses were made of a composite material made of macro wooden structures (principal and secondary posts fixed with wattle) covered by clay mixed with straws and chaff, this allowed a good air-draught during the firing of the house, from a micro- to a macro-level. Despite the air-draught present inside prehistoric houses, experiments support the idea that an accidental fire was not capable of consuming the wattle and daub structures since the wooden

structure was fireproofed with clay (**Fig. 18**). To reach the high temperature required to produce the combustion of the wooden and vegetal inner structure of the walls, a very strong air-draught was needed and this could have been possible during strong atmospheric turbulences; when the protective coating of the daub cracked due to the high temperatures reached, the wooden and the vegetal structures started the ignition process of the walls (**Fig. 19**).

Fig. 18: The wooden structure of the replica of the Chalcolithic house covered with clay mixed with chaff and straws. Vadastra campaign 2003. (Builder Alexandru Stoian, Vadastra village)

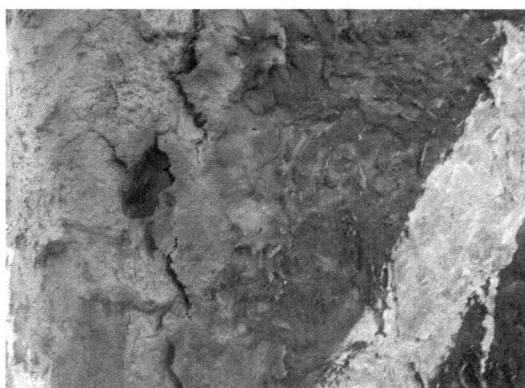

Fig. 19: Combustion of a post inside the wall. One can notice the cracks in the daub protection, the embers inside the perforation produced by heat and (on the right) the unfired coating layer of clay mixed with dung and straws. Vadastra 2006 experiments.

A frequent find in the archaeological record are the ceramic imprints of the wooden inner structure under the shape of tubes or grooves (**Fig. 20**) which survive the processes of dissolution and washing of the fired material of the house.

The process of carbonization and consumption of the wooden structure was accelerated by the collapse of the wooden roof when the extremities of the posts and beams

Fig. 20: Fired wall of a replica of a prehistoric house. Cucuteni 2004 experiments. Conducted by Lecturer Vasile Cotiuga and Dr. Romeo Dumitrescu, with the assistance of the author. One can notice the ceramic tubes resulted after the combustion of the wooden material.

Experiments also demonstrate that generally temperatures over 900°C are difficult to achieve in large structures of combustion, but could be quite easily reached when the walls of the house collapsed creating long narrow tunnels (Gheorghiu 2002a : 84) together with perforated clay pieces from the walls. Similar to parts of the perforated platform in contact with temperatures over 900°C, produced due to strong air turbulence (**Figs 23-24**), the perforated pieces of wall with voids present a change in colour (Munsell 10R/3/6), and the presence of slag and vitreous materials along the channels or tubes, due to the rise of temperature in these narrow voids which change the mechanical properties of these features. Such difference of resistance in the pieces of fired daub allowed a good preservation of the material around the voids, in the shape of tubes or grooves, after the rest of the wall was eroded by weathering.

were in direct contact with air (**Figs 21-22**), and, later, by the (intentional) collapse of the walls. Particularly the collapse of the walls produced large pieces of ceramics with cylindrical voids that raised the temperature sometimes over 1000°C due to the air-draught created.

Figs. 23-24: Fragments of a perforated kiln platform fired at 1150°C. Vadastra Campaign 2002. One can notice the difference in the firing (as colour and texture) between the inner surface of the ceramic tube and the body of material.

After quenching with soil the collapsed walls, the oxygen contained in the micro voids of the material which was in contact with fire, initiated a migration to the exterior of the iron ions existing in clay (V. Burghelea, *personal communication*), a process that transforms in time the natural grey colour (Munsell 7.5YR/7/6) of the unfired material into orange-red (Munsell 10R/6/8) even in the places not fired to become ceramic. This explains the large quantities of red coloured crushed material present in the fired houses not always being the result of a direct firing, this material not being real ceramics.

Combustion in ceramic objects

When discussing South East European Chalcolithic material culture, one can observe that the "second products revolution" (Sherratt 1981) signified at the same

Figs. 21-22: A corner post during the combustion process and after. Vadastra 2006 experiments. The imprint in the clay formed a ceramic groove after the consuming of the wood.

time a cultural isomorphism between food preparation and heating, as well as a change (or specialization) in objects' design that produced a change at the social level; a comprehensible example for the design change are the pyroinstruments in which fire is locked and regulated by means of straining, like the alimentary products, and for the social change is the emergence of a new kind of *oikos* as a complex object designed to dry, to heat, to boil, to bake, to keep embers burning, to smoke, to start fire, to evacuate smoke, and probably more.

Perceived from an integrative perspective Chalcolithic air-draught pyroinstruments could be characterized as an ensemble made of objects-functions-symbols that are in a holistic relationship of function and meaning and which define the living space. Therefore the household could be perceived as a place of sieving and modelling fire and flour, which could be transformed in an instrument of combustion.

In analyzing all the air-draught pyro-objects in a household, one can see that the relationship between them is not only functional and temporal, but also rhetorical. This assumption is sustained by Chalcolithic art representations as vases or architectural models that are in a rhetorical relationship: the house is perceived as being a brazier and vice versa, the oven is perceived as being a house and vice versa, the oven is perceived as being a brazier and vice versa. So the analysis of the objects in relation to fire should not individualize things, but put into evidence their relationships, because fire should be approached as a whole in relationship to the household.

Even if they form a whole, there is a differentiation between the objects that are in a close relationship with fire, so in this perspective, there are many kinds of fires. There are weak fires that need to be protected or regenerated, bright fires for heating, temperate fires for cooking, slow fires for smoking, dying fires for blackening ceramics. With one exception, at the end of the house's life cycle, all these fires were sieved, the ceramic pyro-objects being designed to filter fire.

Photographs: D. Gheorghiu

Acknowledgements

I owe many thanks to Professor Dorin Theodorescu, Olt County Department of the Ministry of Culture, who financed the 2003 and 2004 campaign of experiments in Vadastra. Thanks also to Dr. Romeo Dumitrescu who invited me to fire a replica of a Cucuteni house in 2004 and financed the building of the fired house in Vadastra, to Professor Virginia Burghelea for the information about the chemical transformations in fired clay and to Bogdan Capruciu, MA, for the reviewing of the translation of the text. I would also like to thank Dr. Wendy Logue.

Many thanks to Catalin Oancea, Marius Stroe, Dragos Manea and Stefan Ungureanu who built the house for experiments in Vadastra.

The Project Vadastra of pyroexperimentation was financed by the World Bank and CNCSIS (grant 112) in 2000-2002 in 2003-2004 partly by a CNCSIS grant (No. 1612) and in 2006 by a CNCSIS grant (No. 945).

Bibliography

ARNOLD, E. D.
  1997 *Ceramic theory and cultural process*, Cambridge, New York, Port Chester, Melbourne, Sydney: Cambridge University Press.
BUDIS, M.
  1998 *Microcosmosul gospodaresc. Practici magice si religioase de aparare*. Bucharest: Paideia.
COMSA, E.
  1976 Caracteristicile si insemnatatea cuptoarelor de ars din aria culturii Cucuteni-Ariusd, *SCIVA* 27(1): 23-33.
COMSA, E.
  1990 Complexul neolitic de la Radovanu, *Cultura si Civilizatie la Dunarea de Jos*, VII, Calarasi.
CONTENSON, H. de, and L.C. COURTOIS
  1979 A propos des vases en chaux: Recherches sur leur fabrication et leur origine, *Paléorient* 5: 177-182.
DECA, E.
  1982 Centrul de ceramica din comuna Lungesti, Jud. Valcea, *Buridava – Studii si materiale*, Ramnicu Valcea Museum, Ramnicu Valcea, pp.211-216.
DUMITRESCU, H.
  1968 Un modele de sanctuaire decouvert dans la station eneolithique de Cascioarele, *Dacia* NS, XII: 381-394.
DUMITRESCU, V., H. DUMITRESCU, M. PETRESCU-DIMBOVITA and N. GOSTAR
  1954 *Habasesti*, Bucharest: Editura Academiei.
ELLIS, L.
  1986 Culture contact and culture change during the Copper Age north of the Danube, in *The Bronze Age in the Thracian lands and beyond*, Milan: Dragan European Foundation.
GHEORGHIU, D.
  2002a Fire and air-draght: Experimenting Chalcolithic pyrotechnologies, in Gheorghiu, D. (ed.), *Fire in Archaeology*, BAR International Series 1089 Oxford: BAR Publishing: 83-94.
  2002b On Palisades, Houses, Vases and Miniatures: the Formative Processes and Metaphors of Chalcolithic Tells, pp.93-117 in Gibson, A. (ed.), *Behind Wooden Walls: Neolithic Palisaded Enclosures in Europe*, BAR International Series 1013 Oxford: BAR Publishing.

2003 Water, tells and textures: A Multiscalar approach to Gumelnita hydrostrategies, pp. 39-56 in Gheorghiu, D. (ed.), *Chalcolithic and Early Bronze Age Hydrostrategies*, BAR International Series 1123 Oxford: BAR Publishing.

GOURDIN, W.H. and W.D.KINGERY
1975 The Beginnings of Pyrotechnology: Neolithic and Egyptian lime plaster, *Journal of Field Archaeology* 2: 133-150

HAMER, F.
1975 *The Potter's Dictionary of materials ad technologies*, London: Pitman Publishing; New York: Watson Guptill Publications.

KINGERY, W. D.
1997 Operational Principles of Ceramic kilns, in Rice, M. (ed.), *The Prehistory and History of Ceramic Kilns*, Proceedings of the Prehistory and History of Ceramic Kilns, 98th Annual Meeting of the American Ceramic Society in Indianapolis, pp.11-20.

LAZAROVICI, Gh. and Z. MAXIM
1995 *Gura Baciului. Monografie arheologica*, Cluj-Napoca.

LICHARDUS, J., A. FOL, L. GETOV, F. BERTHEMES, R. ECHT, R. KATINCAROV and I. KRASTEV ILIEV
1996 *Bericht ueber die bulgarish-deutchen Aussgrabungen in Drama (1989-1995)*, Mainz am Rhein, Phillip von Zabern.

MARINESCU BILCU, S., D. POPOVICI, G. TROHANI and R. ANDREESCU
1997 Archaeological researches at Bordusani-Popina (1993-1994), in *Cercetari Arheologice*, 10: 65-69.

MARKEVIC, V. I.
1981 *Pozne-Tripolskie plemena severnog moldovii*, Kishinev.

MAXIM, Z.
1999 *Neo-Eneoliticul din Transilvania*, Bibliotheca Musei Napocensis, Cluj-Napoca.

RYE, O.S. and C. EVANS
1976 *Traditional pottery techniques of Pakistan:*

Fields and laboratory studies, Smithsonian Contributions to Anthropology, No. 21.

SHEPARD, A. O.
1956 *Ceramics for the archaeologist*, Carnegie Institution of Washington, Publication 609.

SHERRATT, A.
1981 Plough and pastoralism: aspects of the secondary products revolution, in Renfrew, C. and S. Shennan (eds.), *Ranking, resource and exchange*, Cambridge: Cambridge University Press, pp.13-26.

SIMSON, J.
1997 Prehistoric ceramics in Mesopotamia, in Freestone, I. and D. Gaimster (eds.), *Pottery in the making. World Ceramic Traditions*, London: British Museum Press.

SERBANESCU, D.
1997 Modele de locuire si sanctuare eneolitice, *Cultura si civilizatie la Dunarea de Jos* XV: 232-239.

STEFANOVIC, M.
2002 Burned Houses in the Neolithic of Southeastern Europe, in Gheorghiu, D. (ed.), *Fire in Archaeology*, BAR International Series 1098, Oxford: BAR Publishing, pp. 55-62.

TRINGHAM, R.
1992 Households with faces: The Challenge of gender in prehistoric architectural remains, pp.93-131 in Gero, J. and M. Conkey (eds.), *Engendering Archaeology. Women in Prehistory*. Blackwell.

VITELLI, K.
1995 Pots, potters, and the shaping of Greek Neolithic society, in Barnett, W. K., and Hoopes, J. W. (eds.), *The Emergence of Pottery. Technology and innovation in ancient societies*, Washington and London: Smithsonian Institution Press, pp. 55-63.

WOOD, J.
1999 Bunsen burners or cheese mould? A new reinterpretation of a Bronze Age ceramic, *Abstracts*, 5th EAA Meeting, Bournemouth.

A Re-Interpretation of a Bronze Age Ceramic.
Was it a Cheese Mould or a Bunsen Burner?

Jacqui Wood

In the summer of 1998 I was working at Lake Ledro pile dwelling museum in northern Italy and was invited to examine a collection of ceramic objects in their exhibition. The museum director wanted me to make replicas for a new exhibit. Three of these objects were categorised as 'sieves of conical bellied form, described as milk boilers or cheese moulds' (Tomasi 1982: 19). On closer inspection of the moulds, it was evident that they had been subjected to repeated contact with intense heat. The inside of one of these perforated pots was almost vitrified, taking on a similar appearance to ceramic crucibles used in bronze smelting. Initially I thought it could possibly have been a type of lantern top, used out of doors during windy conditions. During the next few days I made replicas of these objects in order to ascertain if this was a valid assumption. In order to do this I would need a fuel source for the lantern top.

During the last ten years I have conducted research into the instrumentality of fire in all aspects of prehistoric settlement activities. One such research thread was into different lighting technologies in prehistoric Europe. I considered that a fat or wax soaked rush fire would be the most effective fuel for these ceramic anomalies. The use of rush lights during the last century is well documented in Britain. The soft rush or *Juncus* was peeled of almost all of its outer skin leaving a thin strip of green to give it stability. This was dipped into fat or wax and left to dry. Bundles of these rushes were clipped together to form primitive candles. However, I discovered that if the rushes were peeled entirely, then dipped in fat or wax a more versatile light could be made. If a small light was required an incense type of pot could be filled with the dipped rushes to make a small yet mobile bright light. If however a bright light was needed a large bowl could be filled with rushes and the ensuing light can be as bright as a car headlight. The useful applications of this type of bright light to a prehistoric culture are obvious. The dipped rushes can be made in large quantities and stored until needed. I found that it takes one and a half-hours to produce 15 g of white pith. Fifteen grams of pith dipped into either beef fat, lamb fat or beeswax weighs 350 g (**Fig. 1**).

At an open day at the museum at Lake Ledro I had time to test the lantern top theory. Bundles of wax-soaked rushes were piled onto a plate and lit. I placed one of the 'cheese moulds' on top and it displayed a quite extraordinary effect. A tall pencil like flame rose from the centre of the pot in much the same way as a Bunsen

Fig. 1: *Junkus* fuel.

burner does in a laboratory. The other ceramic was then tested and the same effect was displayed the flame rose to 20 cm in height. Air rushed in through the holes in the sides to produce this controlled flame. The implications of this device were obvious, a controllable flame to be used on a bench is an essential requirement for a jeweller. It could be used for soldering fine metalwork such as gold and silver.

During my work later that year in Poland I noticed at Biskupin in the Bronze Age exhibition, an almost identical ceramic described again as a cheese mould. After working in Poland, I travelled to the EAA conference in Göteborg only to find another mould in the Swedish archaeology (Lunqvist and Ahrberg 1997: 97). Enquires with colleagues at the conference revealed that, similar pots have been found in Europe, in Hungary, Bulgaria, Bavaria and Lithuania. There was a suggestion that some analysis on the pot from Hungary revealed traces of animal fats, but as far as I know this is unpublished. In addition, the use of beeswax as a lighting fuel has been discovered in analysis of the fabric of lamps found in late Minoan Crete (Evershed, Vaughan, Dudd and Soles 1997: 979-85).

On returning to Britain, an article about this discovery was published in British Archaeology in November and consequently I discovered a similar ceramic in the Archaeology of Sussex from the Bronze Age period at Bow Hill. Described thus' This type of vessel may have been used for pressing the whey out of curd in making cheese' (Curwen 1937: 183). I also discovered another much larger example from the Saxon period displayed in the Ashmolean Museum, Oxford.

The following experiments have been undertaken (**Fig. 2**).

Fig. 2: Two ledro burners working.

LEDRO 1
[Size = 8.5 cm high, top diameter 3 cm and bottom diameter 8cm]

Primarily I tested the rush capacity of the pot, in this case, it held 25 g of fat or wax soaked rush. When filled to capacity the rushes were lit from the top. The flame from this pot reached 20 cm high, but after about five minutes, the flame was finding it difficult to sustain itself, due to the packed capacity of rushes. This was inhibiting the airflow from the bottom holes and the flame started to almost blow itself out. I decided to raise the pot on three broken shards and found the flame returned with increased velocity. The flame became wider measuring 3cm across and the burning time of the 25 g of rushes was 15 minutes.

LEDRO 2
[Size = 10 cm high, top diameter 4 cm and bottom diameter 10 cm.]

This pot had a capacity the same as the previous one, 25 g. It was filled and again after about five minutes needed to be raised on potsherds to help the airflow. This flame was however 12 cm longer attaining 32 cm in height and the burning time was 16 minutes.

LEDRO 3
[Size = 8 cm high, top diameter 3 cm and bottom diameter 10 cm]

This pot was quite different to the other two it had a narrow layer of holes at the bottom and the rest of the pot was like a funnel or chimney shape. In addition, because the top was like a narrow funnel and it was not possible to top fill with more than 15 g of soaked rushes. The same problem of airflow was encountered and the pot was raised after five minutes. The height of the flame in this case was 25 cm and the width was an even 3cm. The burning time was 13 minutes.

LEDRO 4
[Size = 7 cm high, top diameter 1.5 cm and bottom diameter 6.5 cm]

This strange funnel shaped object was tested in the same way, the capacity was only 5 g. When the rushes were lit and the funnel was placed on the top, the flame did not come out of the top. The fire went out after smoking profusely. Another possible use was found for this funnel in a later experiment. (**Fig. 3**)

Fig. 3: Funnel Burner.

BISKUPIN/GÖTEBORG
[Size = 11cm high, top diameter 3 cm and bottom diameter 10 cm]

This pot is so named because they are almost identical in size and shape. The top fill capacity was 35 g, the flame height was an efficient 30 cm, and the burning time was 25 minutes. When the flame is not required, a small piece of stone can be place on top of the pot. The flame will just sustain itself enough to keep alight until it is needed again. When the stone is removed and the flame returns. (**Fig. 4**)

SUSSEX
[Size = 7 cm high, top diameter 3.5 cm and bottom diameter 10 cm]

Although this pot did not contain, the amount of holes the others had it performed just as well. The capacity was 20 g and the height of the flame after being raised on shards was 32 cm with a burning time of 15 minutes. Although the burning times were noted in my experiments, because of being raised on potsherds it was quite simple to feed the fire from the bottom during use to extend the burning almost indefinitely.

Fig. 4: Burner with slate on top.

Uses in metallurgy

The prime use that I could initially envisage for this burning device was for soldering metals. However, how would I be able to suspend the object for soldering without the use of a mesh-covered tripod that is used in laboratories? Originally, I thought the device could be placed between two large stones and a metal plate for soldering on it could be suspended between them. This however, would have been insufficient for working on smaller items. While I was trying to solve this problem and think of an alternative in the archaeology to the mesh topped tripod, I recalled another fragment of ceramic from Ledro with 1 cm wide holes in it. It was a flat piece but could possibly have been a fragment of a large round burner. However, the formation of the holes on the underside was interesting. When reconstructing these devices the pots had to be made initially, and then the holes pushed through from the outside with various pieces of stick. The holes on this piece looked like they had been made on a flat surface, and not through the walls of a pot. It occurred to me that perhaps this piece was from another type of device. It could have been used to form a ceramic grill suspended over the burner resting on the stones while soldering is in progress. Hence I made a copy of it and it worked wonderfully. The burning time was the same as the flame filtered thorough the grill making a hot soldering area to work on.

I am not a jeweller, but I found it easy to solder on top of the grill with the burner underneath. (**Fig. 5**)

It occurred to me that this device could also be used for enamelling if a metal object was put on the burner and the Ledro 4 funnel placed on the top to make a sort of furnace oven. When the funnel was tried on top of a piece of metal over the grill, it seemed to draw the hot air through it, without affecting the flames. However, considerable further experimentation is still needed in this area. The applications of these devices could have had many diverse uses in prehistory.

Fig. 5: Soldering metal.

Cheese moulds

Lastly, I will examine the possibility of using these ceramic anomalies for straining curd prior to cheesemaking. A simple curd was made using cheese rennet in much the same way as it would have been made in prehistory. The pots were lined with muslin, in the archaeology lose weave linen would probably have been used for this purpose. The curd was spooned into the pots that were placed on wooden platters to contain the whey. In all cases with the exception of the Sussex pot, the whey only filtered out of the bottom, and not through the holes themselves, thereby making the need for the holes obsolete for cheese straining. (**Fig. 6**) Although the Sussex pot did let some whey run out of the bottom holes, the majority came out of the base.

Fig. 6: No good as cheese moulds.

Conclusion

These devices make mediocre cheese moulds, but very efficient bench fires. This discovery substantiates my theory that only a holistic approach to experimental archaeology will really open doors to our understanding of prehistoric practices. If I had not conducted the research into lighting, I would not have had the fuel to make the burners work.

Bibliography

CURWEN, E. C.
1937 *The Archaeology of Sussex*. London: Methuen & Co Ltd.

EVERSHED, R. P., S. J. VAUGHAN, S. N. DUDD and
J. S. SOLES
    1997 Fuel for thought? Beeswax in Lamps and
conical cups from Late Minoan Crete.
Antiquity 71: 979-85.

LUNQVIST L. and E. AHRBERG
    1997 'Met Kunglig Utsikt' *Arkeologiska Resultat uv
vast Rapport Riksantikvarieambetet*
Goteborge Sweden: 97

TOMASI, G.
    1982 The Lake-dwellings of Lake Ledro. *Natura
Alpina. Rivista della societa di scienze naturali
del trentino* (Trento) 33: 19.

Chalcolithic Copper Source-material and End-products: Early Trade between Israel and Jordan

Sariel Shalev

Introduction

The idea of this paper is to present the results of the archaeometallurgical research concerning the earliest exchange pattern between the source of copper in Faynan, Jordan and the Chalcolithic villages in Israel, where copper products and remains of the use of fire for their production were found.

The socio-economical connections between Jordan and Israel during the 5th and 4th Millennia BC is for a long time a subject of interest to almost all scholars involved in excavations and archaeometallurgical analyses of the finds (Perrot 1955; Bar Adon 1963; Key 1963; Tylecote *et al.* 1974; Notis *et al.* 1984; Levy 1987 and on; Levy and Shalev 1989; Hanbury Tenison 1986; Shalev and Northover 1987 and on; Gilead 1988; Gilead and Rosen 1992; Moorey 1988; Khalil 1989; Khalil and Riederer 1998; Iilan and Sebbane 1989; Tadmor 1989; Tadmor *et al.* 1995; Rostoker *et al.* 1989; Hauptmann 1989 and on; Adams 1991; Adams and Genz 1995; Gale 1991) (For current state of research and bibliography see: Golden *et al* 2001). The aim of this presentation is to analyze the exchange model from the end-products distribution and the archaeological remains of the use of fire for their production in the Chalcolithic villages of southern Israel.

For this purpose I will present the state of research concerning the group of unalloyed copper cutting and piercing implements found in Chalcolithic sites in Israel (Shalev 1994; 1995). The general and overall picture will be presented through a systematic analysis of a single cutting implement found recently in a rescue excavation in Ein Assawir. This unalloyed copper tool will be then compared with 36 published and yet unpublished cutting implements such as flat axes, adzes, and chisels, creating with the addition of more than 33 piercing points such as awls, drills, and hooks, the whole known unalloyed copper products of the Chalcolithic period in Israel.

The production method of these copper artefacts will be reconstructed, based upon metallographic analyses and hardness tests. The source material for the production will be discussed based upon elemental qualitative and quantitative composition analyses (AAS, WDS, EDS). The optional function of these finds will be addressed with the help of macro surface analyses, ethno-archaeological parallels and some preliminary experimental archaeology.

This processed data will be then used for addressing the questions of exchange and economic complexity concerning the ore selection in Wadi Faynan, the transportation of the minerals to the villages on the banks of the Beer Sheva valley in the Northern Negev and the production of these copper artefacts by the use of fire.

The find

During the new excavations in the Chalcolithic village of Ein Assawir, directed by E. Yanaai, a metal blade was found on the floor of house no. 2010 (find no. 70261 from locus 2045), with 6 similar flint axes/adzes. The metal blade was submitted by the excavators for archaeological, metallurgical and metallographic analysis to the archaeometallurgical laboratory at the centre for archaeological sciences, Weizmann Institute of Science.

Typology

The blade is flat and relatively thick (**Fig. 1**). It weighs 284 grams and its dimensions are: 95.5 mm long, 31.4 mm width at its cutting edge, and 18 mm maximum thickness in the centre of its edge. Its section is lens shaped. It has a flat thick butt and a blade merely crescentic which protrude only slightly beyond the body.

The blade from Ein Assawir is part of a group of more than 36 cutting implements such as flat axes, adzes, and chisels (Shalev 1994: 633; 1995: 113, 114) from Chalcolithic villages on the banks of the Beer Sheva Valley (4 at Beer Zafad; 2 at Nave Noy; 5 at Shiqmim), Caves of the Judean Desert (14 at the hoard from Nahal Mishmar; 6 at Nahal Makuch; 1 at Nahal Zeelim), sites near the coastal Plain (1 at Nahal Lachish), and the Jordan valley (3 at Tulelat Ghasul). To these finds we could now add several more finds from the Chalcolithic villages in the Shephela (Shoham and Ein Assawir) and a burial site in the Gallilee (cave of Pekeen).

The object from Ein Assawir fits very well within the typological shape and dimensions of the Chalcolithic blades, ranging from 329 mm to 69 mm long, 45 - 13 mm width at the cutting edge, 23 - 3 mm maximum thickness, and 807 - 80 grams of weight. From those measurements it is quite clear that the major variable factor is the thickness (ratio: 1-7.6) in comparison to the other measurements

Fig. 1: The adze from Ein Assawir (find no. 70261 from locus 2045).

(length ratio: 1-4.8; width ratio: 1-3.5). These variables especially that of the thickness, affect the range in weight (ratio: 1-10.1).

Metallurgical analysis

The metal blade from Ein Assawir was analysed by the author in the archaeometallurgical lab at the Centre for Archaeological Sciences, Weizmann Institute of Science. The surface macro analyses were conducted using an Olympus ZS40 zoom stereoscope. Metallography of a polished sample of the object was examined, un-etched and etched, by Olympus PME3 inverted metallurgical microscope, under incident and polarised light, with magnification of up to 1000. Micro hardness tests, measuring the resistance of the metal sample to indentation, were conducted on a Shimadzu HMV-2000 with a Vickers diamond shaped indenter. The qualitative and quantitative chemical analyses were performed by detecting the spectra and the intensity of the emitted X-ray photons during excitation of the sample surface with high energy electron beam. The elemental analysis were conducted, with the help of E. Klein, on a JEOL 6400 scanning electron microscope (SEM), using energy dispersive spectrometer (EDS) ISIS 300. For the metallographic and the elemental analysis, as well as for the micro hardness testing, a minute chip was cut from the edge of the blade (**Fig. 1**) with a jeweller piercing saw. The sample was hot mounted in a phenolic resin with carbon filler, and then ground and polished for analysis.

Analytical results

The macro surface analysis under a zoom stereoscope presents a significant evidence for the tool production and use: all surfaces show singes of forging. The crescentic blade was shaped, thinned and sharpened by hammering as is evident by the small folded rims of excess material on both edges of the blade. One face of the object is perfectly preserved showing very thin longitudinal cracks created probably by surface stress and tension during use. The other face shows a 'stain' of corrosion that starts circa 20 mm from the butt, continues for 65 mm and ending circa 10 mm from the cutting edge. The corrosion is continuing, on a slightly smaller scale, on the right edge. This 'stain' of corrosion could well be the result of contact with decomposing organic material such as wood. Small, wavy like and cup like depressions are observed on the straight face-edge rims. These edge depressions are visible for less than 20 mm long, starting circa 5 mm from the butt, on the edges of the intact face and circa 30 mm from the butt at the edges of the corroded face. These marks could well be the result of string or wire friction. The back edge between the corroded face and the butt show some uneven squeezing marks as if it was pressured by an angled wooden handle. The compilation of all the macro analysis results show that this metal object was hammered after being cast, especially in the area of its cutting edge. The object was probably used as an adze, when the cutting edge is horizontal and in a straight angle to the wooden handle. The handle itself was probably carved in the shape of the letter Z without its bottom line. The metal blade was attached to the inner side of the upper angle with strings. A wooden handle of this shape with leather straps still attached to it, but without the blade, was found in a Chalcolithic stratum in a cave in wadi Murabbaat, at the Judean Desert (de Vaux 1961: 20, 4). There are no signs of significant weariness on the cutting edge, due to short time of usage and/or use on soft materials with low resistance, like carving wood. In a small experiment that was conducted in the field outside my lab, I tried to observe the effect of carving pine and olive trees with this tool. The results show that it is possible and very easy to use this tool for peeling and carving wood. The macro analysis of the cutting edge before and after use shows no signs of deformation or wariness.

The elemental analysis shows that this Chalcolithic tool was made from unalloyed copper with impurities of less than 0.5Wt%, below the detection limit of the EDS system. These results are in accordance with the analyses of tens more Chalcolithic blades (see for example: Shalev 1991; Hauptmann 1989: Notis *et al.* 1984; Key 1980). According to the results of the analysis of 18 additional Chalcolithic blades that were performed by the author, they are all made of unalloyed copper, identical in composition to the copper piercing points and totally different from the high alloyed coppers with antimony or nickel and arsenic, of the prestige/cult objects of the same period. Out of the 18 analyzed blades, 5 had more than 0.1% arsenic (As) and in 5 others more than 0.1% nickel (Ni) with the arsenic. All impurities concentrations are not acceding 0.5%. The same pattern of impurities distribution and concentration was detected in the analyses of the copper points. These unalloyed copper blades and points, including the one from Ein Assawir, were probably produced in the Chalcolithic villages along the banks of the Beer-Sheva valley, and then distributed further north to other Chalcolithic sites (Shalev 1991; 1994; 1995). To produce these tools, high grade copper minerals were brought to the Northern Negev from Feinan, the nearest copper bearing ore deposits (i.e. Hauptmann 2000), where the mineral copper would have been picked and sorted (Shalev 1991). Several kilograms of ore, mainly cuprite, have been found on settlement sites in the Beer Sheva valley. The pieces of ore are relatively small and do not usually exceed 10 mm in length and 50 grams in weight. They were found alongside grinding and firing installations (Perrot 1955: 79; Shalev and Northover 1987: 362) within the habitation remains of the Northern Negev villages. Slag and crucible remains were associated with the ore at these sites, and the copper prills in the slag were identical in composition to the copper tools from the same sites.

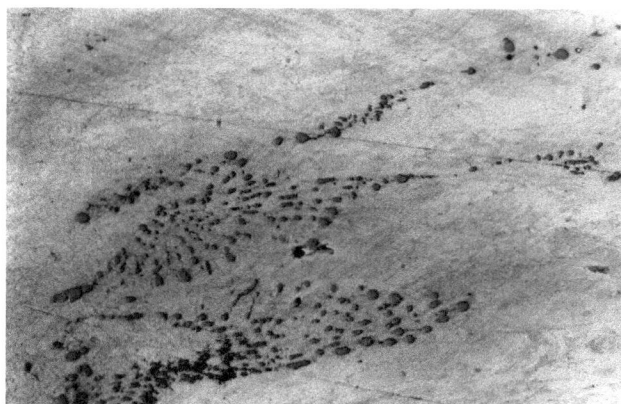

Fig. 2: Metallography of the adze from Ein Assawir x200 (for 36 mil frame), un-etched.

The metallographic analysis shows an original structure of relatively big copper grains of circa 0.15 mm (=150 micron), the boundaries between the original as cast grains is marked by rows of small inclusions, mainly of copper oxides. From the original cast structure we can conclude that the blade from Ein Assawir was cooled relatively slowly and therefore was probably cast in a clay or sand mould. The presence of copper oxides (**Fig. 2**) presents oxidation of the melt. The distribution pattern of the oxides reflects oxygen absorption by the molten copper probably during the pouring process and during solidification (in the grain boundaries. Therefore we could conclude that the copper for the Ein Assawir blade was poured into an open clay or sand mould, probably in one of the Beer Sheva villages, were it was also smelted from ore that was brought to the site from Feinan. The sample, taken from the side of the cutting edge, show after being etched, a typical wavy structure (**Fig. 3**) of heavy hammering of the cast in the cutting edge area. These cycles of cold working, annealing and final cold working is becoming clearer with higher magnification (**Fig. 4**).

Fig. 3: Metallography of the adze from Ein Assawir x50 (for 36 mil frame), etched.

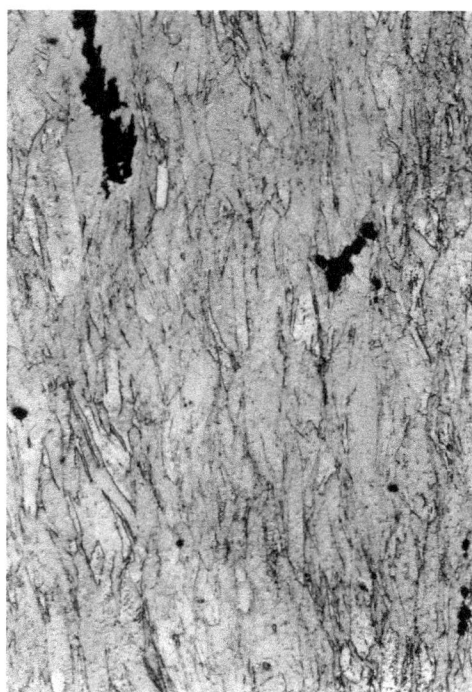

Fig. 4: Metallography of the adze from Ein Assawir x200 (for 36 mil frame), etched.

The equiaxed grains are totally bent and distorted reaching the size of 100 micron X 10 micron, with a ratio of 1:10. The twinning is preserving the annealed stages of circa 500 degrees centigrade for several minutes each time and the slip traces are the evident of the final cold working of the blade.

An identical metallographic picture was found in the microstructure of 6 other Chalcolithic blades from Nahal Makuch and Nahal Zeelim, analyzed by the author. All these blades were at least partially recrystallised. Parts in the middle of the blade, between the surface and the edge, were left annealed with un-deformed equiaxed grains and twinning with no signs of final cold working. Other parts, of the butt and the cutting edge, show a significant deformation of the grains and slip traces. We measured a total reduction of 40% in the butt area and of more than 50% in the cutting edge zone.

The maximum micro-hardness that was measured on the cutting edge of the blade from Ein Assawir is 131Hv. By comparing these measurements to other samples from different areas of other Chalcolithic blades we could see that the cutting edge was made to be the hardest part of the object. The other hard area was the butt (up to 115Hv) and the main body of the blade, even very close to the surface, was left relatively soft, giving hardness measurements of 89Hv to 84Hv. Even this relative soft area is circa twice the hardness of an as cast copper Shalev 1996). Therefore, the entire blade was hammered and annealed. The faces and the edges were hammered and left annealed and twice as hard as the original cast. The working areas, especially the cutting edge, were hammered to achieve a tool hard and as resistible blade as possible in its working areas and softer but still hard enough in all other outer parts. The Chalcolithic metal smiths knew very well the properties of their metal, how to use fire for extracting it from the copper mineral and how to achieve by selective forging the outmost from this unalloyed metal. It would take, at least, another Millennium in the history of human technology of this region, until the hardness of metal objects would be improved by alloying the copper first with arsenic – reaching 204Hv – and later with tin – reaching 250Hv – in the cutting edge zone.

Summary

The story of ancient copper production and utilisation during the first known use of metals by humans, in the Chalcolithic period, in Israel and Jordan, is now – after slightly less than half a century of uncovering the finds – much clearer. The recent intensive archaeo-metallurgical research of the Chalcolithic copper tools from habitation sites, burials and hoards from Israel, and the investigation of the copper minerals from southern Jordan, brought to light the whole copper production chain of the 5th and 4th Millenniums BC in the southern Levant. The results of the last 15 years of research enable us to reconstruct the Chalcolithic copper extraction and production in great details. The story of this production starts at Faynan, Jordan with the identification of very rich self-fluxing copper minerals. This high-grade ore was collected from its original geological occurrence and then, after being carefully sorted, was transported north. From southern Jordan the copper minerals were taken through the Aravah rift, passing Ein-Yahav, to the Chalcolithic villages of the northern Negev, on the banks of the Nahal Beer-Sheva valley. There, whenever needed, with the use of fire on site, these copper minerals were smelted refined and melted, probably in a single simple process, and copper blades and points were made. These copper objects were cast and then hammered and annealed into their final shape and the desirable mechanical properties. Fire was used not only to extract liquid metal out of rock minerals but to manipulate the properties of the end product, as well. The metal objects were then distributed further north and were found abandoned in habitation sites, deposited in burials and collected and kept in hoards. Other metals, made of a total different material, in a different production method, ended as products of distinct different shape and colour, were found beside the copper blades and points in the same Chalcolithic sites of Israel. But, as opposed to the copper products, not a single trace of their production residues, or their source material, were found, as yet.

Bibliography

ADAMS, R.
1991 The Wadi Fidan Project, Jordan 1989. *Levant 22*: 181-183.
ADAMS, R. and H. GENZ
1995 Excavations at Wadi Fidan 4. *Palestine Exploration Quarterly 127*: 8-20.
BAR ADON, P.
1963 The Cave of the Treasure (Hebrew). Jerusalem
1980 The Cave of the Treasure. Jerusalem
GALE, N. H.
1991 Metals and Metallurgy in the Chalcolithic Period. *Bulletin of the American School of Oriental Research 282/283*: 37-61.
GILEAD, I.
1988 The Chalcolithic Period in the Levant. *Journal of World Prehistory 2*: 397-443.
GILEAD, I. and S. ROSEN
1992 New Archaeological Evidence for the Beginning of Metallurgy in the Southern Levant. *Institute of Archaeo-Metallurgical Studies Newsletter 18*: London, 11-14.
GOLDEN J., T. E. LEVY and A. HAUPTMANN
2001 Recent discoveries concerning Chalcolithic metallurgy at Shiqmim, Israel. *Journal of Archaeological Science* 28-9: 951-963.
HANBURY TENISON, J. W.
1986 *The Late Chalcolithic to Early Bronze Age I Transition in Palestine and Transjordan.* B.A.R. International Series 311. Oxford.
HAUPTMANN, A.
1989 The Earliest Period of Copper Metallurgy in Feinan/Jordan. In: Hauptmann A., E. Pernicka

and G.A. Wagner (eds.) *Old World Archaeometallurgy*: 119-135.

2000 Zur fruhen Metallurgie des Kupfers in Fenan/Jordanien. Bochum.

IILAN, O. and M. SEBBANE
1989 Copper Metallurgy, Trade and the Urbanization of Southern Canaan in the Chalcolithic and the Early Bronze Age. In: de-Miroschedji P. (ed.) *L'urbanisation de la Palestine a L'age du Bronze ancien*: B.A.R. International Series 527(1): 139-162. Oxford.

KEY, C. A.
1963 The trace-element composition of the copper and copper alloys of the Nahal Mishmar hoard. In: Bar Adon P. (ed.) *The Cave of the Treasure*: 238-243 (Hebrew). Jerusalem.

1980 The trace-element composition of the copper and copper alloys of the Nahal Mishmar hoard. In: Bar Adon P. (ed.) *The Cave of the Treasure*: 238-243. Jerusalem.

KHALIL, L.
1989 Excavation at Magass-Aqaba 1985. *Dirasat 15(7)*: 71-109.

KHALIL, L. and J. RIEDERER
1998 Examination of Copper Metallurgical Remains from a Chalcolithic Site at el-Magass, *Jordanian Damaszener Mitt. 10*: 1-9.

LEVY, T. E. (ed.)
1987 *Shiqmim I*. B.A.R. International Series 365. Oxford.

LEVY, T. E.
1995 Cult, Metallurgy, and Rank Societies – Chalcolithic Period. In: Levy T.E. (ed.) *The Archaeology of Society in the Holy Land*: 226-244. New York.

LEVY, T. E. and S. SHALEV
1989 Prehistoric Metalworking in the Southern Levant. *World Archaeology 20*: 350-375.

MOOREY, P. R. S.
1988 The Chalcolithic Hoard from Nahal Mishmar, Israel, in Context. *World Archaeology 20(2)*: 171-189.

NOTIS, M. R., H. MOYER, M. A. BARNISIN and D. CLEMENS
1984 Microprobe Analysis of Early Copper Artifacts from the Northern Sinai and the Judean Caves. In: Romig Jr. A. D. and J. I. Goldstein (eds.) *Microbeam Analysis - 1984*: 240-242. San Francisco.

PERROT, J.
1955 The Excavations at Tell Abu Matar, near Beersheba. *Israel Exploration Journal 5*: 17-41; 73-84; 167-189.

ROSTOKER, W., V. C. PIGOTT and J. R. DVORAK
1989 Direct Reduction to Copper Metal by Oxide-Sulfide Mineral Interaction. *Archeomaterials 3(1)*: 69-87.

SHALEV, S.
1991 Two different copper industries in the Chalcolithic culture of Israel. In: Mohen J-P. and Eluere C. (eds.) *Decouverte du Metal*: 413-424. Paris.

1994 The change in metal production from the Chalcolithic period to the Early Bronze Age in Israel and Jordan. *Antiquity 68*: 630-637.

1995 Metals in Ancient Israel: Archaeological Interpretation of Chemical Analysis. *Israel Journal of Chemistry 35*: 109-116.

1996 Archaeometallurgy in Israel: the impact of the material on the choice of shape, size and color of ancient products. In: Demirci S., A. M. Ozer and G. D. Summers (eds.) *Archaeometry 94*: 11-15. Ankara.

SHALEV, S. and J. P. NORTHOVER
1987 Chalcolithic Metal and Metalworking from Shiqmim. In: Levy T. E. (ed.) *Shiqmim I*. B.A.R. International Series 365: 351-371.

TADMOR, M.
1989 The Judean Desert Treasure from Nahal Mishmar – A Chalcolithic Traders' Hoard? In: Leonard Jr. A. and B. B. Williams (eds.) *Essays in Ancient Civilization Presented to Helene J. Kantor*: 249-261. Chicago.

TADMOR M., D. KEDEM, F. BEGEMANN, A. HAUPTMANN, E. PERNICKA and S. SCHMITT STRECKER
1995 The Nahal Mishmar Hoard from the Judean Desert: Technology, Composition, and Provenance. *Atiqot 27*: 95-148.

TYLECOTE, R. F., B. ROTHENBERG and A. LPU
1974 The Examination of Metallurgical Material from Abu Matar, Israel. *Historical Metallurgy 8(1)*: 32-34.

de VAUX, R.
1961 Archaeologie. In: Benoit P., J. T. Milic and R. de Vaux (eds.) *Discoveries in the Judean Desert II, Les Grottes de Murabba'at*: 3-51. Oxford.

Iron Production in the Northern Eurasian Bronze Age

Stanislav A. Grigoriev

Abstract

The earliest iron objects are known in the Near East. They are dated to the 5th – 2nd millennia BC. In Northern Eurasia there are objects made of iron in burials of the Pit-Grave culture (3rd millennium BC) and in complexes of the Late Bronze Age (2nd millennium BC). Some archaeologists suppose that this iron had meteoritic provenance. Usually it is explained by the presence of nickel in the iron objects. However, other scientists suppose that this iron was produced as a result of the copper ore smelting. Our investigations of slag from Northern Eurasia confirmed this opinion. Usually production of small pieces of iron was possible when metallurgists smelted chalcopyrite. During this process wustite was produced, whose reduction to iron took place then. The presence of nickel in many objects in the Near East may be explained by additions of minerals containing Ni and As to the ore. The main purpose of this operation was the production of As-bronzes.

Key words: Bronze Age, iron production, metallurgy, Near East, Northern Eurasia

Introduction

The origin of iron metallurgy is one of the most important problems in archaeometallurgical studies. In the Near East the earliest iron objects are known from the 5th millennium BC (Sammara in Mesopotamia). Then, in the 4th – 2nd millennia the number of iron objects gradually increased, but their number is insignificant, especially against the background of copper objects. We know only 74 iron objects in the Near East dated to the Late Bronze Age (Waldbaum 1980: 69-77). Therefore, during all this period iron was of great value and its finds are very rare. Iron was 35-40 times more expensive than silver. Usually, ornaments and cultic objects were produced of this metal. Therefore, the finds of iron objects are connected, as a rule, with a prestigious context. A dagger from the "royal" tomb at Alaca Höyük had a gold-covered hilt. We know from the Hittite sources a royal throne made of iron. Two daggers with iron blade have been found in the tomb of Tutankhamun. At last, the Hittite king Hattušili III wrote to Shalmaneser about his present – an iron blade (Ivanov 1983: 91-96; Muhly 1980: 37, 50; Waldbaum 1980: 76). In addition, up to the 12th century traces of smelting iron ores are lacking. The mass distribution of iron objects is dated to the same period, and they are already presented not only by ornaments and small objects, but also by relatively large weapons. Thus, the qualitative changes in the iron production took place in this period. The observation of J.D. Muhly, is very interesting, who wrote that the widespread diffusion of the iron in the Near East coincides with the migration of the Dorians and the "Sea peoples" (Muhly 1980: 51).

Such a situation allows a problem to be formulated: what was a reason of these changes, and what provoked such a limited iron production in the previous periods? In some works it is supposed that iron in the ancient Near East was extracted from iron ores, but metallurgists could not regulate this process. This resulted in high cost of this metal and its comparative rarity. However, a purposeful extraction of iron is doubtful. The iron ores are widespread more widely, than copper ores. Therefore, if the technology of smelting iron existed in the 3rd millennium BC (or even earlier), it would become popular rather fast, but we cannot observe this up to the 12th century BC.

Another point of view on the nature of ancient iron is that it had a meteoritic provenance. It may be confirmed by the Hittite texts, where iron is designated as a "metal of heaven", as well as by analyses of ancient objects, which demonstrate frequently a high nickel content, what is viewed as a mark of the meteoritic provenance (Waldbaum 1980). However, many ancient objects contain no nickel. In addition, there are clear descriptions allowing us to suppose that the iron production was an artificial and rather complicated process.

Therefore, the majority of scientists assumes a considerable part of the meteoritic iron among the ancient objects of the Near East, but supposes, that there was any technology of extracting iron from the iron ores, whose nature is unclear now (Waldbaum 1980: 80, 88). J. A. Charles believes that ancient iron was extracted by means of smelting copper ores with the use of ferriferous fluxes. As the temperatures increased and the balance $CO/CO2$ changed, the production of iron grew (Charles 1980: 166; Charles 1992: 24, 25).

I think that slag samples of the Late Bronze Age of the Volga-Ural area allow the nature of this technology to be better understood. In this period the technology of copper smelting started to be changed.

Traditions of copper production in Northern Eurasia

In the Middle Bronze Age metallurgy was based, mainly, on smelting oxidized ores, infrequently secondary sulphides from the ultra-basic rock (Grigoriev 2000a: 517; 2000b: 142). The volumes of smelting operations were rather small, and the smelting atmosphere was reducing. Metallurgical installations were presented by rather small cupola-shaped furnaces, which were often attached to a well. This allowed an additional blowing to be executed because of the difference of temperatures in the furnace and the well.

Metallurgical furnaces have been found in each dwelling of the Sintashta culture of this period. The formation of the culture was stimulated by the coming of population from Southeastern Anatolia (Grigoriev 2002a; 2002b). It is worth to be mentioned, that smelting technology, types of ore, forms and microstructures of slag of this culture have also parallels in this area. In addition, a two-stepped technology of extracting silver from leaden ores is reconstructed, which has analogies in Anatolia too.

At the beginning of the Late Bronze Age the technologies changed. Metallurgy was present in all areas, where ore could be found. In some areas we see the use of new types of ore. In the steppe area, the Sintashta tradition of smelting oxidized ores remained. But more typical became the use of ores from silicate rocks and sandstones. It demanded to increase temperatures and intensify blast, in conditions of absence of tradition to use the ferriferous fluxes. In a number of instances, for example in the Atasu settlement in Central Kazakhstan, it resulted in the considerable metal losses in the form of cuprite (Grigoriev 2003). The metallurgical furnaces of this period, despite formal differences, prolonged the Sintashta tradition. On sites of Central Kazakhstan the furnaces attached to the well, identical to those in the Sintashta culture, have been excavated. On other sites furnaces in pit of round or rectangular form have been found. Rectangular form was caused by the use of stone slabs for lining the walls of furnaces. Such constructions and technological traditions were typical of sites of the Timber-Grave and Alakul cultures, which had been formed on the Sintashta base (Grigoriev 2002a: 138-147).

Metallurgical slag containing a great number of cuprite inclusions is not typical of all sites of these cultures. It may be met in those sites where oxidized ores were used. It is a usual situation on some sites of Central Kazakhstan (Atasu, Myrzhik, etc.) and in the area of the Kargali mines (Grigoriev 2003; Rovira 1999). If sulphide ores were used, slag is not so rich in cuprite. It must also be mentioned that natural sulphide minerals are present in slag not so often being dissociated in metallurgical reactions. But there are in many samples isotropic copper sulphides forming in these reactions.

At the same time, two new cultural traditions appeared and, accordingly, new technologies of copper extraction. They are presented in a new generation of sites connected culturally with the Seima-Turbino and Fyodorovka cultural types, as well as with the later cultures formed on their base. Broadly speaking, these are the sites of Western Siberia, formerly dated to the Early Bronze Age of the area (Odinovo-Krokhalevo, Vishnyovka, Krotovo), sites of the Fyodorovka culture itself in the Urals and Kazakhstan, Cherkaskul and Mezhovskaya cultures in the Urals, and Suskan-Lebyazhinskaya culture in the Volga-Oka area (Grigoriev 2002a: 192-272) Many sites left by these new populations, are situated in the steppe zone, but the main areas of their distribution is the forest and forest-steppe zone. These traditions interplay with the former. Therefore, there is a series of mixed types. Accordingly, new metallurgical technologies can be presented also on sites referred to the former cultural traditions.

At this time a new type of furnaces appeared, which allowed slag to be tapped (**Fig. 1**).

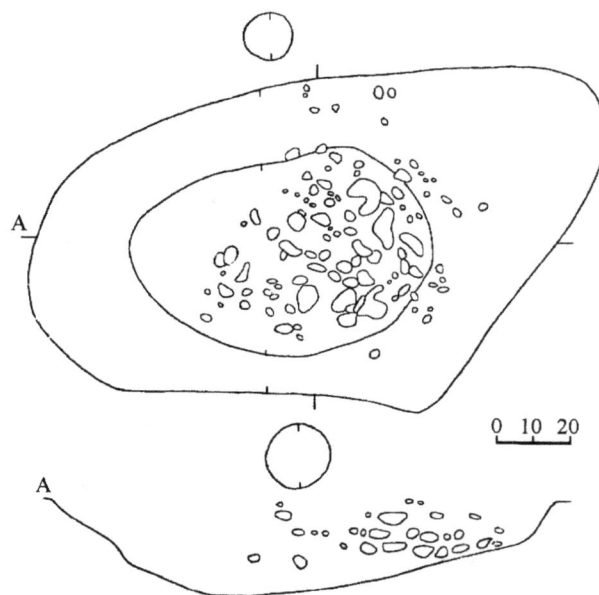

Fig. 1: The furnace from the settlement of Semiozerki (after Evdokimov and Grigoriev, 1996).

At the same time, on many sites where slag has been found, metallurgical furnaces are absent, but fragments of crucibles are present, in which ore was smelting (**Fig. 2**). Especially many crucibles and their fragments have been found in sites of the Mezhovskaya culture and in the settlement of Mosolovskoe of the Timber-Grave culture in the Don area. The volumes of these crucibles were between 600 and 1350 cm^3 (Obydennov and Shorin 1995: 85; Pryakhin 1996: 60). In contrast to the earlier crucible smelting, the sulphide ores were using.

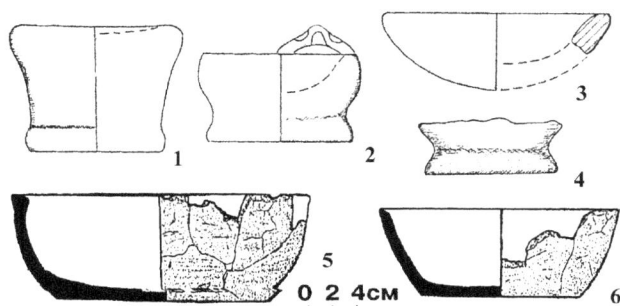

Fig. 2: Smelting crucibles of the Ural and Don areas: 1-4 – Tyubyak (Mezhovskaya culture), 5, 6 – Mosolovskoe (Timber-Grave culture) (after Pryakhin, 1996; Obydennov and Shorin, 1995).

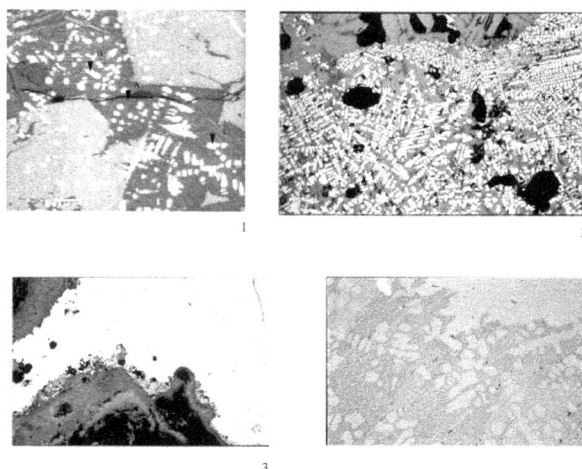

Fig. 3: Microstructures of the Late Bronze Age slag: 1 – lattice structures of wustite (white) in the sample of slag (№236) from the settlement of Ust-Kenetai in Central Kazakhstan, 2 – lattice structures of wustite (white) in the sample of slag (№463) from the settlement of Yukalekulevskoe in the Western Urals, 3 – Iron (white) in the sample of slag (№66) from the settlement of Verkhnyaya Alabuga in Western Siberia, 4 – Cast structures of iron (white) in the sample of slag (№66) from the settlement of Verkhnyaya Alabuga in Western Siberia.

Thus, both technologies used were applied to smelt sulphide ore. In some slag samples small inclusions of chalcopyrite have been detected. Especially interesting for the subject discussed are inclusions of wustite, frequently melted, which had been formed as a result of melting of copper sulphide sinking to the bottom of a furnace or crucible. They indicate a nature of metallurgical reactions. At the first stage the decay of chalcopyrite into copper and iron sulphides took place. The former having a low melting point, melt and drop from the ore pieces. The latter, in the process of burn-out of sulfur and the reaction with oxygen, turn into wustite (FeO). The melting point of this new mineral is lower, that results in the appearance of melted lattice structures, which can be disintegrated into melted dendrites (**Fig. 3.1, 3.2**). In the case of preservation of reducing atmosphere, wustite can turn into iron, albeit not fully, because a considerable part of it, interacting with silicate components, forms fayalite. Thus, quantity of iron extracted from such a smelt, depends not only upon atmosphere, but also upon a ratio of acid and basic oxides in the charge.

In outcome, conditions were created, which allowed a certain quantity of reduced iron to be extracted. Sometimes, it resulted in the formation of alloys copper with iron containing different proportions of these two components. Frequently, it is reflected in the microstructure of the slag, where small inclusions of copper are present, which have optical characteristics intervening between the iron and the copper. E. N. Chernikh also wrote about the use of copper, which was smelted of chalcopyrite, basing on iron admixtures in metal artifacts from the hoard at Sosnovaya Maza (Chernikh 1970: 19).

Production of iron in the Bronze Age

Investigation of metallurgical furnaces in Mitterberg (Austria) has allowed a conclusion to be drawn that in the Middle Bronze Age (that corresponds to the Late Bronze Age in the east) metallurgists smelted chalcopyrite. After such smelts, a thin interlayer enriched with iron formed

on a surface of copper (Preßlinger and Eibner 1987: 237-239). However, in conditions of a crucible smelting of sulphide ores, the formation of iron proceeded, probably, more often, because the more reducing atmosphere formed, whose preservation was promoted by the walls of the crucible, in addition to sulphur connecting with a part of oxygen.

In some slag samples of Northern Eurasia small globules of metallic iron have been found. This does not indicate that the melting point of iron was reached. Most likely, the iron was reduced from the globules of wustite, having a low melting point. A sample from Verkhnyaya Alabuga in the Tobol basin is most interesting in this sense. It is presented by a piece of iron about 3 cm in diameter having a dendritic anisotropic structure[1]. It is necessary to say, that iron, as well as other metals, is isotropic, but polished sections demonstrate sometimes the effect of anisotropy. The formation of this structure is connected, probably, with smelting of chalcopyrite. Optical, chemical and SEM analyses have revealed the predominance of iron with admixtures of copper, wustite, silicates and some other components (**Figs 3.3, 4**; Tables 1, 2).

---

[1] A preliminary study of this sample by R. Schwab demonstrated that this iron was carburized, so we can say that it was steel. This study is uncompleted yet. But I think that in case of smelting in a crucible, there could be conditions created, which allowed iron to be partly carburized and it was not a special process of carburization.

Table 1: Chemical analysis of slag (sample 66) from the settlement of Verkhnyaya Alabuga made in the Chemical laboratory of the Chelyabinsk geological expedition.

SiO$_2$	FeO	CaO	Cu	Fe	Fe$_3$O$_4$
2,34%	–	0.44%	1.99%	81.42%	n.d.

Table 2: SEM-analyses of slag from the settlement of Verkhnyaya Alabuga made in the Technical University of Freiberg by means of the Digital Scanning Microsope DSM 960.

Weight per cent			
Analysis	Material	O	Fe
1	iron		100
2	wustite	22.18	77.82
Atomic per cent			
1	iron		100
2	wustite	49.88	50.12

In a smaller quantity iron is present in slag of many other settlements: several sites in the Ayakagitma and Besh-Bulak areas in the south of Central Asia, Ust-Kenetai and Kara-Tyube in Eastern Kazakhstan, Korshunovo, Verkhnyaya Alabuga and Vishnyovka in Western Siberia, Novokizganovo, Yukalekulevskoe, Baigildino, Novobaryatino, Aitovo and Verkhnebikkuzino in the Western Urals, Shigonskoe and Popovo Ozero in the Volga area, and Mosolovskoe in the Don area (**Fig. 4**).

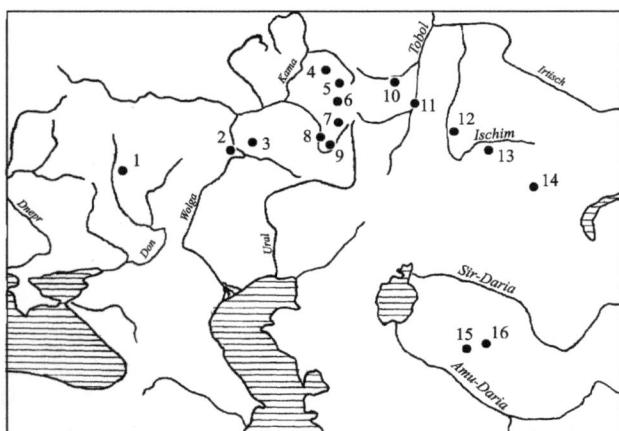

Fig. 4: Finds of the copper slag with iron inclusions: 1 – Mosolovskoe, 2 – Shigonskoe, 3 – Popovo Ozero, 4 – Novokizganovo, 5 – Yukalekulevskoe, 6 – Baigildino, 7 – Novobaryatino, 8 – Aitovo, 9 – Verkhnebikkuzino, 10 – Korshunovo, 11 – Verkhnyaya Alabuga, 12 – Vishnyovka, 13 – Ust-Kenetai, 14 – Kara-Tyube, 15 – Besh-Bulak, 16 – Ayakagitma.

Thus, in the Late Bronze Age of the Volga-Ural region there were conditions to obtain iron as a by-product after smelting sulphide copper ore, which should stimulate searching for evidences of the use of this iron. There are several facts demonstrating it. Fifty years ago B. N. Grakov wrote about the finds of iron objects in the settlements of the Bronze Age (Grakov 1958). Iron objects are also known on sites, in which the slag with iron inclusions is found, for example, the Mosolovskoe settlement (**Fig. 5**) (Pryakhin, 1996: 55). It seems to be very possible, that there were some other such finds, but their stratigraphical position was not so clear.

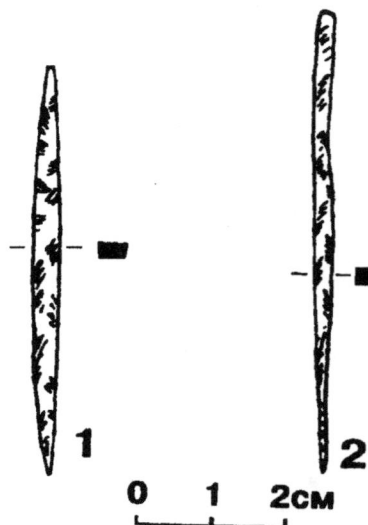

Fig. 5: Iron objects from the settlement of Mosolovskoe (after Pryakhin, 1996).

In the Belozerka period of the Northern Pontic area (12th – 9th centuries BC) the number of iron objects (mainly, daggers and knives) increased. However, it is supposed that they had been imported from the Balkan-Danube area (Nikitenko 1998). And it is very hard to say what does this mean: the appearance of the true iron metallurgy or more intensive exploitation of the sulphide copper ores.

All said above about the Volga-Ural situation is quite applied for the Near East. Probably, a small part of the iron objects in the Near East was actually obtained from the meteorites. This is indicated also by the etymology of words for the designation of iron (Egyptian "metal of heaven" and Hittite "black metal of heaven") (Waldbaum 1980: 79). At the same time, the admixture of nickel (which is typical of many objects in the Near East) cannot be a sign of the meteoritic iron. The nickel-containing iron could be made as a result of smelting copper ore. Theoretically it is impossible, as the nickel in the system Ni-Cu-Fe is connected completely with the copper. But in some instances the nickel may be equally distributed between the iron and the copper (Tylecote 1981: 48). We must remember that metallurgical processes in antiquity were not pure chemical processes and the single processes could be divided in the time and the space of the furnace. In addition, in more composite systems, which include the arsenic, the nickel will be shared between the iron and the copper. Alloying the copper with the arsenic took place in Anatolia and Transcaucasia already in the 4th millennium BC. In my work on the Sintashta metallurgy I have demonstrated that alloys copper with arsenic were obtaining at the stage of ore smelting (Grigoriev 2000a; 2000b). Taking into account the Near Eastern origin of the Sintashta culture (Grigoriev 2002a; 2002b), we may assume the use of a similar way of alloying in the Near East. In this case the ratio of iron objects with nickel and without it does not indicate a ratio between the meteoritic and metallurgical iron, but a ratio between alloyed and non-alloyed metal. The use of nickel-containing bronze in the Near East is known widely enough. High concentrations of nickel in ancient iron objects of this area are sometimes accompanied by high concentrations of arsenic (Piaskowski 1991: tab. 1).

There are also other evidences of obtaining iron as a by-product of smelting copper ores in the Near East.

The earliest iron objects contain admixtures of copper. For example in objects from Tepechik the copper contents was 6.12 and 2.19% (Yener *et al.* 1994: 383).

In the large mining center Timna in the south of Palestine the by-production of copper and iron is detected for furnaces of the 14th – 12th centuries BC. This conclusion is checked by means of study of lead isotopes in copper and iron objects (Gale *et al.* 1990: 182-191). In Alalakh (15th century BC) the alloy of copper with iron is found (Wertime 1980: 15).

Smelting experiments of J. D. Muhly and R. Maddin with chalcopyrite without fluxes resulted in obtaining some metallic iron in slag (Wertime 1980: 16). Experimental works of R. F. Tylecote, who used oxidized copper ores containing significant quantity of iron oxides, were quite successful too (Tylecote 1980: 188, 189).

Some written sources of the late 3rd – early 2nd millennium BC allow us to suppose, that the iron was the slag of copper smelting (Ivanov 1983: 108).

Raw, from which iron was making, is very indicative. In Old Assyrian texts the term *amutum* indicates the iron from the ore. The raw for it was *aši'um*. It is supposed that it was any iron ore, but do not hematite, because it was the object of trade. Most probably, this term indicated any metal (Muhly 1980: 35, 36). It is possible to suspect, it was iron – a product of smelting chalcopyrite, which required the following treatment.

The possibilities for obtaining iron from chalcopyrite existed in the Near East for a long time. On Cyprus the use of sulphides began in the middle of the 4th millennium BC. From the same time arsenic ligatures are also known (Zwicker 1987: 194, 199). Earlier iron objects could be made partly of the meteoritic iron, partly of iron, which had been received of the copper ores with the use of ferriferous fluxes. However, we may not exclude an episodic use of the sulphide ores in this period too.

Of course, such a way of iron production could not guarantee a stable result. This explains a rarity of this metal, and also the mentions about poor and good iron. If oxides or sulphur were unreduced in the iron, the iron lost ductility, although its visual characteristics were not too different.

Against this background the earliest finds of iron objects in Eastern Europe are of special interest. The finds in tombs of the Pit-Grave culture in the Orenburg area, dated to the 3rd millennium BC, are presented by a chisel-shaped tool, a bimetallic adz with a hilt of copper and a working edge of iron, and a disc-shaped object. Microscopic investigations of these objects allowed a conclusion to be drawn that they were made of the meteoritic iron. (Terekhova *et al.* 1997: 33-39). This conclusion quite corresponds to that drown above because sulphide ores are absent in this area, and iron production from copper ores was impossible there, especially in this period.

Conclusions

Thus, the large part of the earliest iron objects of the Near East, including those containing nickel, had not a meteoritic, but metallurgical provenance. There is a basis to think, that it was obtained, mainly, as the by-product of smelting chalcopyrite. Instability of its obtaining was connected with this, as well as its rarity and expensiveness. The widening of the sulphide ores' extraction resulted in a gradual increase of a number of iron objects. However it is not a reason to say about the beginning of the iron metallurgy. Judging from the mentioned above connection of the migrations of the Dorians and the "Sea peoples" with the distribution of iron in the Near East, the beginning of the true iron metallurgy took place on the European continent. It is most possible that the Dorian southward movement was stimulated by the drastic movements of the Central European populations, presented by the bearers of the Urnfield culture. This allows us to search for roots of the true iron metallurgy in Central Europe, where the extraction of the sulphide ores was known starting from the Middle Bronze Age, which corresponds to the Late Bronze Age in Eastern Europe (Woelk *et al.* 1998: 263). From the same period the iron objects appear in this area (Pleiner 1980: 378), which also corresponds to the situation in the Volga-Ural area. Their rarity allows us to think that they had been obtained from the smelting chalcopyrite too. Probably, it was a common situation when the use of the sulphide ores and metallurgy with slag foregone to the appearance of iron (Craddock 1999: 175).

I have already wrote that the essential cultural transformation in Central and Western Europe at the end of the Early Bronze Age – beginning of the Middle Bronze Age was conditioned by the coming of the Volga-Ural populations (Grigoriev 2002a: 213-222, 259-267). Probably, just these tribes introduced skills of extraction of iron from the suplphide copper ores everywhere in Northern Eurasia.

Such an approach to iron distribution can be confirmed by the appearance in Central Europe of bronze objects made according to the Seima-Turbino tradition and the Andronovo-like ceramics on settlements of Southern Germany. It is also possible to suppose that the following migrations of populations from Europe resulted in the rise of the iron metallurgy in the Near East.

However, the situation needs even more complicated explanation. At the beginning of the Late Bronze Age (I remind that it corresponds to the beginning of the Middle Bronze Age in Central Europe) in the Urals and Western Siberia the cultures appeared whose origins were connected with the Near East. In this period, tin-bronzes and tools and weapons with cast socket appeared here in addition to some other cultural features (Grigoriev 2002a: 206-211). So, it is quite possible that this technology of ore smelting came in the area together with these other features in material culture.

In contrast to this, R. Pleiner believes that the skills of iron production penetrated Europe in the 2nd millennium BC immediately from the Near East through Greece, Balkans and the Caucasus (Pleiner 1980). However, this conclusion cannot be confirmed by other materials.

Thus, the essential technological changes, which took place in the Volga-Ural metallurgy in the beginning of the Late Bronze Age, were stimulated by impulses from the Near East. These impulses introduced the primary knowledge of iron. Then the Volga-Ural area played an essential role in the distribution of this knowledge.

Acknowledgements

I would like to express my heartfelt gratitude to people who put at my disposal materials which allowed me to do this work: E.N. Chernikh, T.M. Potyomkina, G.B. Zdanovich, M.V. Vokhmintsev, A.D. Pryakhin, A.V. Vinogradov, V.S. Gorbunov, M.F. Obydennov, O.V. Kuzmina, Yu.I. Kolev. A special thank to E. Pernicka, who helped me to study some samples by means of SEM-analyses in Freiberg.

Bibliography

CHARLES, J. A.
1980 The Coming of Copper and Copper-Base Alloys and Iron: A Metallurgical Sequence. In Werime, T. A. and J. D. Muhly (eds.) *The Coming of the Age of Iron*. New Haven, London: Yale University Press, pp. 151-181.
1992 Determinative Mineralogy and the Origins of Metallurgy. In Graddock, P. T. and M. J. Hughes (eds.) Furnaces and Smelting Technology in Antiquity. *British Museum Occassional Paper* 48: 21-28

CHERNIKH, E. N.
1970 *Drevnyaya metallurgiya Urala i Povolzhia*. Moscow: Nauka ("Ancient metallurgy in the Urals and Volga Region").

CRADDOCK, P. T.
1999 Paradigms of metallurgical innovation in prehistoric Europe. In Hauptmann, A., E. Pernicka, T. Rehren and Ü. Yalcin, (eds.) *The beginnings of metallurgy*. Bochum: Deutschen Bergbaumuseum Bochum, № 84 (Der Anschnitt: Beiheft: 9), pp. 175-192.

EVDOKIMOV, V. V. and S. A. GRIGORIEV
1996 Metallurgicheskie kompleksy poseleniya Semiozyorki II. In Grigoriev, S. A. (ed.) *Novoe v archeologii Yuzhnogo Urala* ("Metallurgical komplexes of the settlement of Semiozerki II". New in archaeology of the Southern Urals). pp. 124-130.

GALE, N. H., H. G. BACHMANN, B. ROTHENBERG, Z. A. STOS-GALE and R. F. TYLECOTE
1990 The Adventitious Production of Iron in the Smelting of Copper. In Rothenberg, B. (ed.) *The Ancient Metallurgy of Copper*. London: Institute for Archaeo-Metallurgical Studies. pp. 182-191.

GRAKOV, B. N.
1958 Stareishie nakhodki zheleznykh veshei v evropeiskoi chasti terrotorii SSSR ("The earliest finds of iron objects in the European part of the territory of the USSR"). *Sovetskaya arkheologia* 4: 3-12.

GRIGORIEV, S. A.
2000a Metallurgicheskoe proizvodstvo na Yuzhnom Urale v epokhu sredney bronzy ("Metallurgical production of the Middle Bronze Age in the Southern Urals."). *Drevnyaya istoriya Yuzhnogo Zauralia* (Ancient history of the Southern Urals) Chelyabinsk: Rifei: 444-531.
2000b Investigation of Bronze Age Metallurgical Slag. In Davis-Kimball, J., E. M. Murphy, L. Koryakova and L. T. Yablonsky (eds.) *Kurgans, Ritual Sites, and Settlements Eurasian Bronze and Iron Age. British Archaeological Reports*. International Series 890. pp. 141-149.
2002a *Ancient Indo-Europeans*. Chelyabinsk: Rifei.

2002b The Sintashta Culture and the Indo-European Problem. In Jones-Bley, K. and D. G. Zdanovich (eds.). *Complex Societies of Central Eurasia from the 3rd to the 1st Millennium BC. Journal of Indo-European Studies Monograph Series* 45. Wasington D.C.: Institute for the Study of Man. pp. 148-160.

2003 Metallurgia epokhi bronzy Zentralnogo Kazakhstana. In Akishev, K. A., V. F. Varfolomeev, V. F. Zaybert and M. K. Khabdulina (eds.) *Stepnaya zivilizacia Vostochnoi Evrazii.* Astana: Kyultegin. ("Metallurgy of the Bronze Age of Central Kazakhstan." In Steppe civilization of Eastern Eurasia). pp. 136-158.

IVANOV, V. V.
1983 *Istoria slavyanskikh I balkanskikh nazvanii metallov.* Moscow: Nauka ("History of the Slavic and Balkan names of metals").

MUHLY, J. D.
1980 The Bronze Age Setting. In Werime, T. A. and J. D. Muhly (eds.) *The Coming of the Age of Iron.* New Haven, London: Yale University Press. pp. 25-67.

NIKITENKO, N. I.
1998 Nachalo osvoeniya zheleza v belozerskoi kulture, in *Rossiiskaya archaeologia,* 3 ("The beginnings of the use of iron in the Belozerka culture").

OBYDENNOV, M. F. and A. F. SHORIN
1995 *Archaeologicheskie kultury pozdnego bronzovogo veka drevnikh uraltcev.* Ekaterinburg: Ural University ("Archaeological cultures of the Ural Bronze Age (Cherkaskul and Mezhovskaya cultures)").

PIASKOWSKI, J.
1991 Ancient metallurgy of iron in the Near East. In Wartke, R. B. (ed.) *Handwerk und Technologie im Alten Orient. Internationale Tagung.* Berlin: Philip von Zabern. pp. 75-83.

PLEINER, R.
1980 Early Iron Metallurgy in Europe. In Werime, T. A. and J. D. Muhly (eds.) *The Coming of the Age of Iron.* New Haven, London: Yale University Press. pp. 375-416.

PRESSLINGER, H. and C. EIBNER
1987 Bronzezeitliche Kupferverhüttung im Paltental. In Hauptmann, A., E. Pernicka and G. A. Wagner (eds.) *Archäometallurgie der Alten Welt. Beiträge zum Internationalen Symposium "Old World Archaeometallurgy",* Heidelberg. ("The Bronze Age Copper Smelting in Paltental". Archaeometallurgy of the Ancient World). pp. 235-240.

PRYAKHIN, A. D.
1996 *Mosolovskoe poselenie metallurgov-liteyshikov epokhi pozdnei bronzy.* V.2. Voronezh: University press ("The Mosolovskoe settlement of ancient metallurgists of the Bronze Age").

ROVIRA, S.
1999 Una propuesta metodológica para el estudio de la metalurgia prehistórica: el caso de Gorny en la region de Kargaly (Orenburg, Rusia), *Trabajos de prehistoria* 56(2) : 92-111.

TEREKHOVA, N. N., L. S. POZANOVA, V. I. ZAVIALOV and M. M. TOLMACHOVA
1997 *Ocherki po istorii drevnei zhelezoobrabotki v Vostochnoi Evrope.* Moscow: Metallurgia ("Studies on History of Ancient Iron-working in Eastern Europe").

TYLECOTE, R. F.
1980 Furnaces, Crucibles, and Slags. In Werime, T. A. and J. D. Muhly (eds.) *The Coming of the Age of Iron.* New Haven, London: Yale University Press. pp. 183-228.

1981 Chalcolithic metallurgy in the Eastern Mediterranean. In: Reade, J. (ed.) *Chalcolithic Cyprus and Western Asia. British Museum Occasional Paper* No 26: 41-51

WALDBAUM, J. C.
1980 The First Archaeological Appearance of Iron and the Transition to the Iron Age. In Werime, T. A. and J. D. Muhly (eds.) *The Coming of the Age of Iron.* New Haven, London: Yale University Press. pp. 69-98.

WERTIME, T. A.
1980 The Pyrotechnologic Background. In Werime, T. A. and J. D. Muhly (eds.) *The Coming of the Age of Iron.* New Haven, London: Yale University Press. pp. 1-24

WOELK, G., P. GELHOIT and W. BUNK
1998 Reconstruction and operation of a Bronze Age copper-reduction furnace. In Rehren, T., A. Hauptmann, and J. D. Muhly (eds.) *Metallurgica Antiqua.* Bochum. pp. 263-278.

YENER, K.A., E. GECKINLI and H. ÖZBAL
1994 A brief survey of Anatolian metallurgy prior to 500 BC. In Demirci, S., A. M. Özer, and G. D. Summers (eds.). *Archaeometry 94. The Proceedings of the 29th International Symposium on Archaeometry.* Ankara. pp. 375-391.

ZWICKER, U.
1987 Untersuchungen zu Herstellung von Kupfer und Kupferlegierungen im Bereich des östlichen Mittelmeeres (3500-1000 v. Chr.). In Hauptmann, A., E. Pernicka, and G. A. Wagner (eds.) *Archäometallurgie der Alten Welt. Beiträge zum Internationalen Symposium "Old World Archaeometallurgy",* Heidelberg. ("Study of the origins of Copper-alloys in the area of Eastern Mediterranean" Archaeometallurgy of the Ancient World). pp. 191-201.

Pyrotechnology of Titelberg Iron Age Coin Production

Ralph M. Rowlett and Dragana Mladenovic

The University of Missouri excavations on the Iron Age oppidum Titelberg in Southwestern Luxembourg has for the main objectives the clarification of the material and habitational counterparts of the Celtic La Tène culture as described by ancient Classical authors. This research was sponsored by the University of Missouri Research Council and the National Science Foundation.

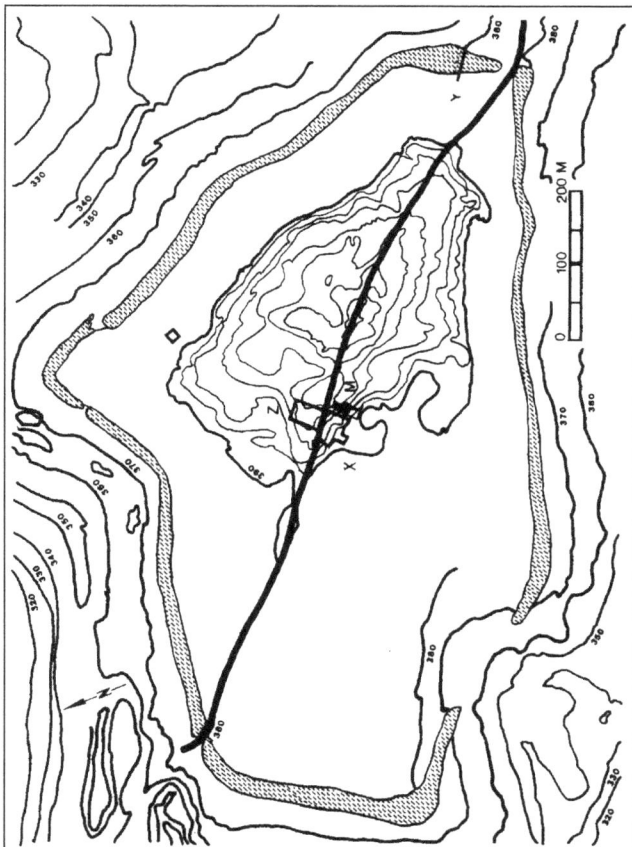

Fig. 1: Map of the Titelberg.

Near the highest point in the center of the Titelberg (**Fig. 1**) the Missouri excavations found stratified 17 house floors of which 13 date to the pre-Roman Iron Age, so our major objectives were met. The excavations took place mainly from 1972-82 with additional field work in 1988 and 1991. During this time archaeologists from the Luxembourg State Museum excavated in other precincts

on the Titelberg (Rowlett 1988; Rowlett, Thomas, and Sander-Jorgensen 1982; Thomas, Rowlett, and Sander-Jorgensen 1975; Thill 1965; Metzler and Weiller 1977; Metzler 1994).

By coincidence, the four most recent pre-Roman layers (ca. 150 BC to 30 BC) were the floors of mints for producing coinage. Coinage continued after the Roman Conquest until about AD 50 through 4 more stratigraphic layers. From these floors and exterior layers some concept of how coins were produced can be obtained. It can be seen how the chiefs of this particular tribe (ostensibly the Treveri) managed their money supply, as this is probably the only mint operation for the Treveri at this particular time. Outside the building were various furnaces and smelters for producing the metal for the coin flans. Although no one stratigraphic layer contained a complete array of the features and artifacts enabling the minting process, from the total set of mint associated floors and layers the entire process can be illustrated.

The locality close to the central apex of the Titelberg where the coins were produced had already enjoyed a privileged status even before coins were minted there, as in each rebuilding of the 17 floor levels, an attempt was made to follow the outlines of the previous building and to superimpose the main fireplace more or less exactly over the earlier ones. At present, it is difficult to discern exactly what was so special about the pre-mint building locale, but once the mint was established, the significance of the locality is more obvious. Perhaps the emplacement gained its elevated status by association with the chiefs of the tribe.

The issuance of coins in the name of the chief as the personification of the tribal polity raises the question as to how closely were the chiefs involved in the actual production of coins. In Roman and medieval times moneyers could produce coins for the rulers without the latter having to be too directly involved. To test this proposition, we reasoned that whether one uses a redistributive or a cybernetic model of chieftaincy, chieftains should have wider contacts than most individuals of their societies and have more access to exotic materials. The sheer number of coins around a chief's building should be greater than for an ordinary structure, and should give an indication of the development of the chieftain's external contacts and/or supplies.

Calculation of the distance from point of origin of the foreign coins found on the Titelberg gives the ranges of foreign coins. Comparison of the ranges of coins associated with the various mint levels on the Titelberg shows that the ranges of coins found in and around the mint precinct exceed that of the Titelberg in general, an independent sample taken from the Luxembourg State Museum's excavations and from surface finds. The steady increase in the ranges of alien coins around the mint buildings shows the growth and extension of the contacts of the mint operators. Since the coin ranges here are greater than for the Titelberg in general, one concludes that the chiefs were more or less directly concerned with mint operations. What is more, the ranges of the alien coins from the pre-mint emplacement being greater than that from the rest of the Titelberg, even in its pre-mint days, implies that already this building was somehow involved in the activities of the chiefs of the Titelberg and its tribe. Later in the early first century, the mint structure was rebuilt and successively refurbished to convey the glory of the Roman way itself.

Mint Foundry Buildings

Coinage begins at the center of the Titelberg during the Pale Brown II floor occupation in early La Tène III or D1 times. The Pale Brown building was a squarish wattle and daub timber house, very much like the pre-mint phase Pale Brown III house. Already a transverse cellar had been established on the south side of the house, a cellar

location which continued until the building of the Stone Foundation House about AD 1. Superimposed above the Pale Brown Ia and Ib (gravel flooring) minting floors is the Orange Clay Floor of a square building. This clay floor was heavily worn when excavated, but must have been originally 2 cm thick. The clay itself was obtained from the deepest subsoil in the uppermost clay stratum of the Titelberg butte itself.

The yellow green clay of the superimposed Green Clay Floor house was extracted from a deeper stratum of clay on the Titelberg. This particular clay floor, also 2 cm thick, was kept particularly clean and neat within its wattle and daub enveloping structure. A well was placed in service on the east side of the building toward the side street. This must have been the thatched; timber framed mint building standing at the time of the Roman Conquest and is tree ring dated to before 30 BC when the Dalles Floor mint was built (**Fig. 2**).

The Dalles Floor mint shows the first signs of Romanization, within a now larger timber framed structure floored by flagstone paving both on the main level as well as in the cellar (**Fig. 3**). In the larger structure, the fireplace was somewhat out of alignment. The cellar had its own fireplace as well as access to a domed bake oven that was certainly used to bake coin moulds, and probably bread as well. During this phase the furnaces for melting the coin metal was moved from the north side of the buildings to the adjacent east of the Dalles Floor building, where a large smelter slightly less

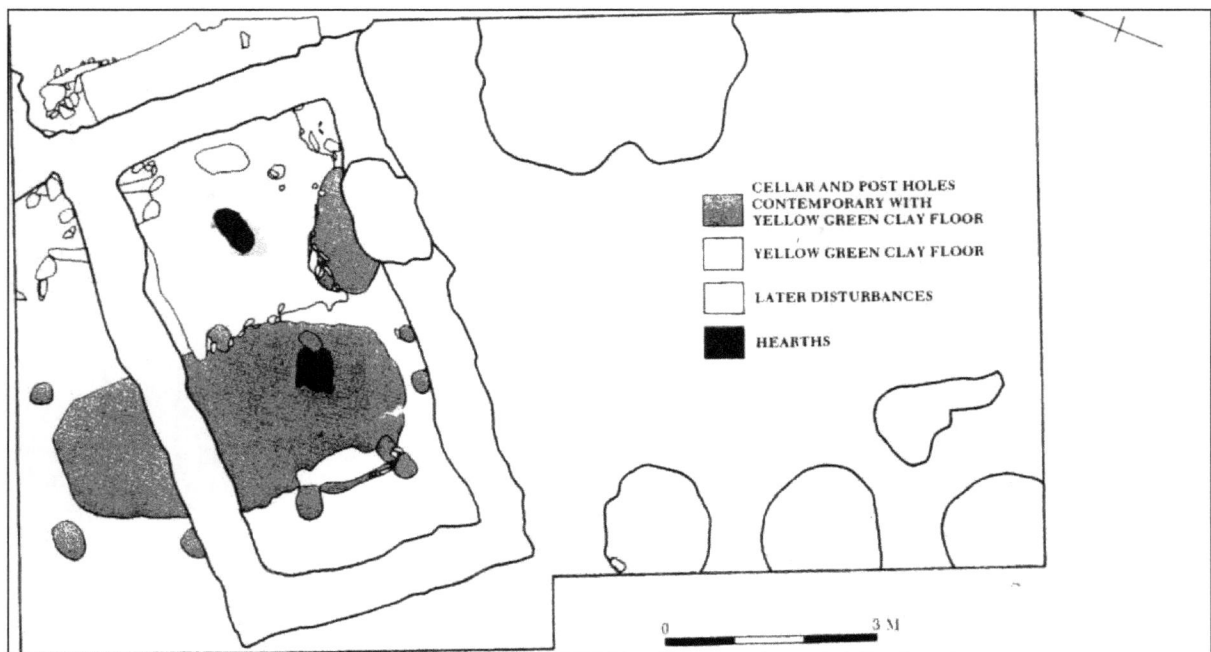

Fig. 2: The Green Clay Floor building and contemporary structures.

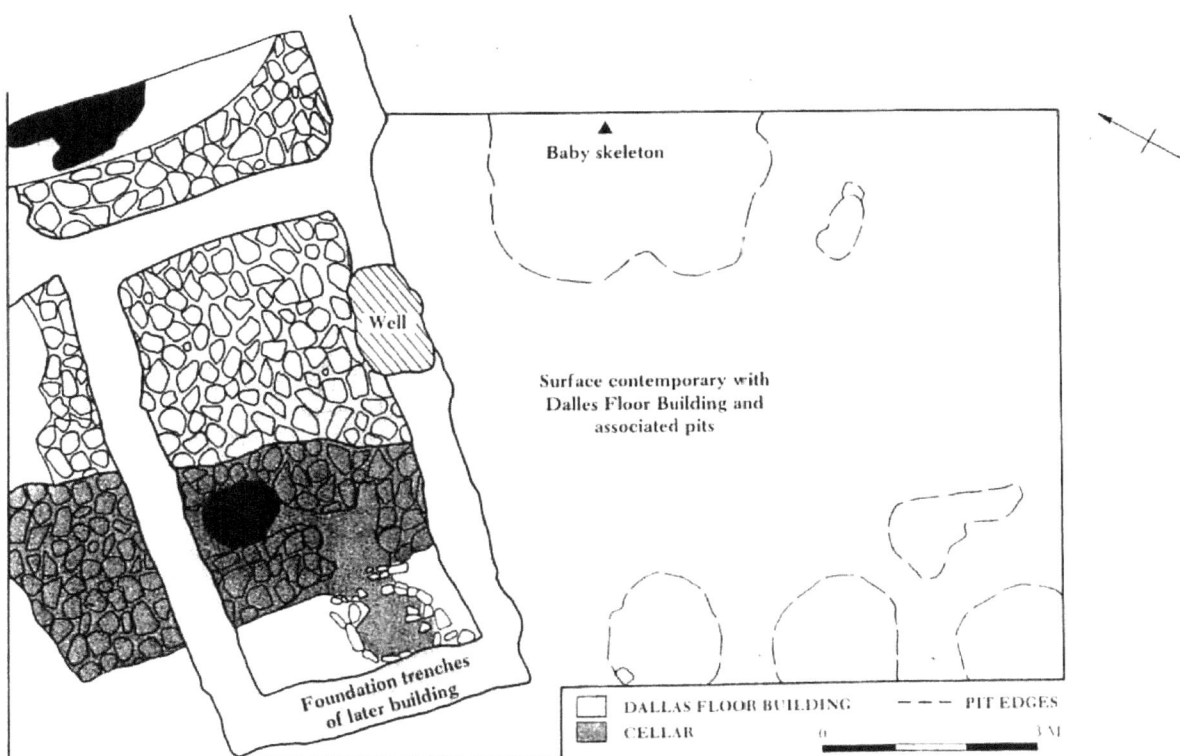

Fig. 3: The Dalles Floor mint foundry.

than 2 x 2 meters was used, through several renovations. This smelter and the well were both rapidly refilled upon the initiation of construction of the succeeding Stone Foundation House mint.

The smelter for the final mint, the Foundation House, has not yet been located, although the well south of the building was excavated by Jean Krier of the Luxembourg state museum. The Foundation House was a considerably Romanized structure.

In the spirit of Roman engineering, the house has stone foundations 130 cm. deep that provide the name for the building. On the foundations stood timber framed and plastered walls which together with stone columns, roofing tiles, glass covered windows, and fern carpeted floors represent a totally new category of building and living standard, unknown at the site before. At first the building consisted of a single room, but just a few years later another chamber, presumably of the same size, was added to the north of it, as well as a small corridor providing access to both rooms connecting them. The result of this enlargement was a long building that stretched parallel to the side street, surrounded firstly by a colonnade of wooden columns and subsequently by stone ones. An idealized reconstruction (**Fig. 4**) shows this building with two floors, which is an inference based on the surviving model houses (*aediculae*) (Thill 1965) made of limestone, and windows, whose presence is

attested by both finds of window glass and the pattern of coin moulds and flans distribution around the house. It seems that the workers in the mint have had the habit of throwing unsuitable moulds and flans through the windows, creating thusly clusters of finds in the radius of each window opening.

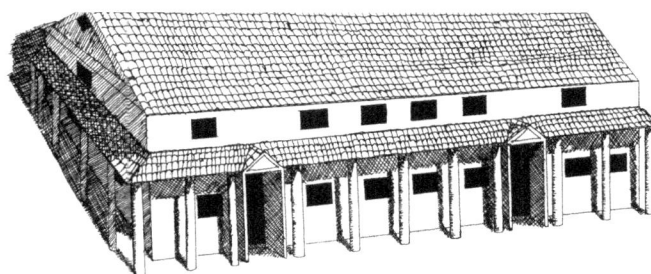

Fig. 4: Idealized Reconstruction of the Foundation House.

The inventory of the mint from this period is better known than from those preceding it. The South Room contained a brazier fireplace at its northern end, above the spot where the fireplaces were previously in all but the Dalles Floor phase. In the south part of the room a working bench made of roofing tiles was compiled. It could have been a metallurgical pouring bench, since many metal (copper alloy) splashes were found on and around it. Other smaller finds fit into a repertoire of objects one would expect to find in a mint: coin moulds, coin die, metal cutters for trimming moulds, touchstones for streak tests, and finally the final products — coins. A slightly unusual find is a little bronze head of the god Taranis lying close to the NE corner of the room. The head is flat at the back with a small tang indicating that it was meant to be mounted. Cult implications aside, and provided that metallurgical analyses substantiate it, the head could represent a local product indicating that the mint occasionally might still have engaged in a wider range of metal objects.

The North Room of the Foundation House is far less preserved due to its having been partly dug out for a 3rd century house cellar and its far northern segment was destroyed by Franks around AD 400 when they constructed a smelter over the north end of where the North Room formerly stood. In the south part of the room stood a stone anvil with cutting marks on it. An assortment of small tools was found here, primarily chisels and hammers to cut apart coins and rouelles cast in series. Some coins and roundels were recovered uncut.

Mints, especially regional ones, were very important for the Romans and their relationships with local populations. The mints represented the state in a much more graphic way than most other institutions. Out of mints came perhaps the only images of the ruling emperors that an ordinary person would ever see. Officials seem to have taken certain steps to make mints look more grand, official, and "Roman" for the subordinates. The replacement of the original wooden columns on the mint building with stone ones might have been such a move. The limestone columns were further plastered to resemble a much more luxurious stone — marble. Within the Foundation House are the traces of other decorative architectural elements, like painted plaster and plaster mouldings. These features might have played a more symbolic and less pragmatic role — to represent Rome, its glory and authority at this distant place.

The Roman style foundations for the last mint building were unwisely run above the old Dalles well and the deepest part of the old, abandoned cellar. This placement was followed by a later subsidence of the foundations and the cracking and deterioration of the Foundation House structure (Rowlett 1997). This contributed eventually to the razing of the last mint structure about AD 50, a date determined from the close correlation to the early Roman ceramic assemblage at St. Albans (Verlamium), England. Whether this razing coincides with any Imperial strife or conflict is not apparent at this time.

The Minting Process

A generalized idea of the minting process can be obtained from the evidence from all these mint floors. Coin flans would be cast in the coin moulds, which had to be prepared from the following steps:

A. Mine blue clay from geological deposits at the base of the Titelberg.

B. Obtain Eifel lava temper for the moulds from ancient volcanoes on the northern edge of the tribal territory.

C. Grind Eifel lava into powder on stone hand querns.

D. Mix Eifel lava powder with blue clay so that the clay will endure high temperature heating.

E. Form flat rectangular moulds ca. 10 x 15 cm and 8 mm thick.

F. Make coin flan depressions by inserting in the moulds a wooden dowel the diameter of the intended coins.

G. Forms for wheel-shaped small coins make by impressing old wheel coin in clay (Dalles Floor).

H. When possible, re-use old coin moulds.

Coin Flan Production

Metal for coin production was presumably obtained from the nearest sources:

A   Copper from own tribal territory in deposits near Fischbach, Germany.

B   Tin imported from Brittany.

C   Silver and lead imported from neighboring tribe to the south, the Mediomatrici near Metz in Lorraine, France.

D   Gold imported from Rhineland?

E   Re-cycling of older metals?

Alloys

Metallic alloys for coinage were combined in smelters, one to the east and one to the north of mint. The resulting metal ingots are represented by the bun-shaped copper-based ingot found sitting on the Green Clay Floor. These ingots, fractured or chisel-cut to suitable size, were crushed into powders with hand querns. (Household querns for grinding bread flour were already of the rotary type, using the Eifel lava from the north-eastern sector of the tribal region.) The crushed or powdered alloys were then placed in the pockets of the ceramic coin moulds to produce disk-like flans when heated beyond the melting temperature of the alloy. Traces of metal, as detected by both x-ray fluorescence and neutron activation indicate that bronze, silver and gold coins were produced concurrently, at least in late La Tène times, (Garrison 1997) and that the coin moulds were re-cycled as long as they could stand up to the heat of repeated firing and the efforts made to dislodge the flans from the moulds.

Designs were cut into the iron die faces by quartz or steel engravers. After the production of the flans, designs were struck on the faces of the re-warmed flans by dies such as the three hammer or "reverse" dies recovered by the Missouri project. Faces of the worn dies were further obliterated to inhibit unauthorized use.

Coins were issued in the name of the tribe and/or chief, such as ARDA. The names on the coins were written in either Greek or Latin letters, or some mixture thereof. This use of script was hardly novel, as writing on bones relating to deities occurs as far back as Middle La Tène on the Titelberg. ARDA's name is actually written more often in an ostensibly retrograde direction than in orthograde. His name might actually have been ADRA, but ARDA, as "High One" makes more sense as a Celtic name.

Coin mould frequency gives quantifiable indications as to the infusion into the money supply of new issues during the course of late La Tène. There is a notable steady increase in coinage until some seemingly critical cutback just before the Roman Conquest, then an increase under the fiction of self-rule until the time of Augustus. Caesar's lieutenant Aulus Hirtius apparently had coins honoring Caesar, with an elephant, issued under his own name from Green Clay floor times. Hirtius' successor Carinas continued this tradition from the Dalles Floor. Under Tiberius and later emperors coin production continued in the now decrepit Foundation House until it was razed about AD 50, at which time the mint was limited to the productions of rouelles or wheel money, an extremely minor denomination.

The finishing of the coin issues is best illustrated in the final, Romanized mint, the Stone Foundation House. The South Room was used primarily for preparing the fragmented metal to put in the coin mould cavities, to judge from the number of fragments and splashes found around the tile bench in the South Room. Flan striking must have taken place in the South Room as well, warming the flans over the brazier fire to soften them for striking.

The North was the scene of finishing the issues. The last issues from there were a set of rouelles, cast in series. These were cut apart on a large cuboid calcareous anvil stone in the center of the North Room. Damaged, mashed rouelles and a small hammer and cutting chisels lay at the sides of the anvil stone.

This emission of rouelles apparently ends the long series of issues at the Titelberg mint foundries location. The re-cyclable bronze rouelle wasters, the serviceable tools, and the Taranis figurine left behind imply a certain haste in the cessation of the issues. Was the decrepit Foundation House in the process of collapsing or did the Romans decide to put an end to this minting associated with tribal independence? Although minting finally ends on the Titelberg, later under Roman Imperial instigation it resumes at nearby Trier, within the same civitates territory (Wightman 1971).

Summary

The Titelberg mint foundry series, occupying eight floor levels in four different superstructures, represents a very unusual archaeological discovery. These mints being all superimposed stratigraphically led to some disturbances, but also made the chronological relationships unequivocal and facilitates direct comparison from one mint to another, making it possible to trace the coinage process and its development and variations. The continuity persisting for virtually a century after conquest into the Gallo-Roman Imperial era enables the drawing upon the better known Roman practices. Most of the steps of coinage are represented at the Titelberg mint foundries, with only the striking hammer, striking anvil and obverse dies not being recovered or recognized.

Pyrotechnology comes into play in five main instances, not counting light for night work or heating for the minters during any winter issues. **First**, there is the offsite pyrotechnology of producing from ores the raw ingots of copper, tin, lead, silver and gold used for making the coins. **Secondly** comes the on-site pyrotechnology for forming the carefully assessed alloys for the coin issues. The simple furnaces had to achieve temperatures of 1,300 degrees for copper to smelt, the other metals having lower smelting points. Such temperatures had to be re-achieved in the casting of the flans. The flan moulds themselves involved a **third** pyrotechnical step. While the initial firing of the moulds could take place in an ordinary bake oven, much like the way modern Pueblo Indians fire both pottery and tortillas simultaneously in the same dome shaped oven. Since the moulds had to be able to withstand the smelting of the powdered or granular alloy, the **fourth** step again demanded temperatures near 1,300 degrees. Lower temperatures, such as could be achieved and tolerated inside the mint buildings, were sufficient for the softening of the flans for striking, the **fifth** pyrotechnical step, and any annealing that needed to take place when the series cast coins were cut apart. Pyrotechnology for the bloomery production of iron and steel for the iron tools comes into consideration as well, and although iron production from the rich local ores took place elsewhere on the Titelberg (Metzler 1994), this iron pyrotechnology was not directly practiced at the locale of the mint foundries.

Table I Coin production sequence on the Titelberg in the late La Tène Iron Age and in early Roman times

I. Manufacture of ceramic moulds

A. Mine blue clay from geological deposits at the base of the Titelberg.

B. Obtain Eifel lava temper for the moulds from ancient volcanoes on the northern edge of the tribal territory.

C. Grind Eifel lava into powder on stone hand querns.

D. Mix Eifel lava powder with blue clay so that the clay will endure high temperature heating.

E. Form flat moulds ca. 10 x 15 cm. and 8 mm. thick.

F. Make coin flan depressions by inserting in the moulds a wooden dowel the diameter of the intended coins.

G. Forms for wheel-shaped small coins make by impressing old wheel coin in clay (Dalles Floor).

H. When possible, re-use old coin moulds.

II. Coin Flan Production

A. Metal for coin production obtained from the nearest sources:

   1. Copper from own tribal territory in deposits near Fischbach, Germany.

   2. Tin imported from Brittany.

   3. Silver and lead imported from neighbouring tribe to south, the Mediomatrici near Metz in Lorraine, France.

   4. Gold imported from Rhineland?

   5. Re-cycling of older metals?

B. Use smelters to produce bun-shaped metal ingots (Green Clay Floor)

C. Grind copper, tin and lead into powder to produce alloys on stone hand querns (Green Clay Floor).

D. Melt metals in smelters north of mint foundry building or over fireplaces within foundries.

E. Grind alloys into powder to put measured amounts into coin mould depressions.

F. Bake coin moulds and coin flans at smelters north of the mint foundry building.

G. Shake flans out of coin moulds that sometimes had to be broken in the process.

(Pale Brown II through Foundation House floors.)

H. Re-cycle still serviceable coin moulds.

I. Cut series-cast flans apart with small "tack" hammer and chisels. (Dalles Floor; North Room of Foundation House.)

III. Striking the Flans

A. Cut designs of coin faces on steel dies with a quartz and/or steel engraver and steel twist drills. (Dies found in Pale Brown and Light Brown floors; quartz gravers in Pale Brown and Orange Clay floors.)

B. Heat coin moulds over fireplaces in foundry to soften metal and place with tongs on anvil die.

C. Strike dies with large steel hammer weighing at least 4.5 kilos (10 pounds). (Inferred from Roman finds and from experimental archaeology.)

D. Remove struck coin and allow to cool while striking other coins.

E. Account record of total coins produced by various coloured small pebbles representing quantities of coins produced.

IV. Putting Coins into Circulation.

A. Payment by chief for goods and services rendered? (hypothetical)

B. Allotment to patri-clan senior member(s) on basis of patri-clan size. (hypothetical)

C. Stage parade by chief and retainers, chief tosses out coins from chariot while parading through villages (historically attested by in Gaul by Classical writers).

Bibliography

GARRISON, E.
   1997 Trace Metals in Coin Moulds from the Titelberg. Abstracts, Society for American Archaeology Meetings, Nashville: Nashville.
METZLER, J.
   1994 *Titelberg, Oppidum Celtique.* Luxembourg: Musée de l'Etat de Luxembourg.
METZLER, J. and R. WEILLER
   1977 *Beitrage zur Archaologie und Numismatik des Titelberges.* Publication de la section Historique de Luxembourg 41: 15-87.
ROWLETT, R. M.
   1997 An Instance of Gaulish Sabotage of Roman Construction. *Abstracts*, Nashville: Society for American Archaeology Meetings, Nashville.

1988 Titelberg, a Celtic Hillfort in Luxembourg. *Expedition* 30: 31-40.

ROWLETT, R. M., H. L. THOMAS, and E. SANDER-JORGENSEN

1982 Stratified Iron Age House Floors on the Titelbourg, Luxembourg. *Journal of Field Archaeology* 9: 301-311.

THILL, G.

1965 *Titelberg*. Luxembourg : Museé de l'Etat de Luxemburg.

THOMAS, H. L., R. M. ROWLETT, and E. SANDER-JORGENSEN

1975 The Titelberg: A Hill Fort of Celtic and Roman Times. *Archaeology* 31: 241-259.

WIGHTMAN, E. M.

1971 *Roman Trier and the Treveri*. Cambridge: Cambridge University Press.

Fire Cult? - The Spatial Organization of a Cooking Pit Site in Scania

Jes Martens

Abstract

Specialized cooking pit sites have attracted some attention during the last decades. They are constituted by a large number of cooking pits either organized in rows or apparently scattered unsystematically around the site. However, even in the latter case it is very seldom that they cut each other. This gives the impression that the often quite large number of pits must have been in use over a rather limited period of time. Unfortunately, the pits are very seldom furnished with anything but fire-cracked stones and charcoal, so the use of the pits is still a matter of discussion. However, research seems to favour an interpretation of the sites as places of cult connected to the use of fire. The present investigation took the outset in these and other assumptions about the cooking pit sites and tried by means of different excavation methods and scientific analyses to get a step further. A major question was: is a cooking pit site just a cooking pit site or are the cooking pits just a symptom of the action? A phosphate mapping gave an especially intriguing hint of the answer to that question.

Introduction

In 1943, Willi Wegewitz was the first who described a specialized cooking pit site by Tangendorf, North Germany. It consisted of 26 disorderly scattered pits filled with fire-cracked stones and soil mixed with charcoal (Wegewitz 1943). The site was apparently isolated in the terrain, the nearest known prehistoric location being a cemetery from the Late Bronze- Early Iron Age. The lack of indications of ordinary settlement activities led Wegewitz to suggest that the pits could be traces after rituals in connection with the cemetery. A similar interpretation was suggested when Rudolf Dehnke 24 years later presented a unilinear system of cooking pits at Bötersen (Dehnke 1968, 1972). The cultic connotation has never since left this type of site, though the number has increased rapidly as has the area of distribution. Today, specialized cooking pit sites are known in a considerable number from Northern Germany, Denmark, Southern Sweden and Norway (**Fig. 1**) (cf. Horst 1985, Henriksen 1999, Martens 2004).

Generally, the sites are divided into two major groups; "ordered sites" at which the pits are arranged in parallel lines, and "disorderly sites" at which there appears to be no system behind the scattering of the pits. However, as the material is increasing so is the variation, and the dividing line between these two major groups is not so obvious anymore. Among other variations are now seen sites at which the pits are arranged in a circle or a circular arc (cf. Narmo 1996, Martens 2004).

The dating of the phenomenon is generally speaking the late Bronze Age and the Early Iron Age (Horst 1985), though the further north one gets, the later the appearance and disappearance (Thörn 1996, Henriksen 1999). Thus in Norway the majority of the sites dates to the Roman Iron Age and the Migration period (Martens 2004).

Fig. 1: The main distribution area of specialized cooking pit sites (prep. by the author).

In 1980, Sigrid Heidelk-Schacht published a study on cooking pit sites from the Northern GDR. From this area she could list 30 such sites and on that basis she tried to point out some general features.

1. The sites are situated at exposed points in the landscape.

2. They are situated close to water.

3. They are "isolated", i.e. there is no archaeologically proven relationship to nearby cemeteries or settlements.

4. They are characterized by a large number of "fireplaces".

5. The "fireplaces" are usually round or oval, rarer are rectangular forms. Their diameter is in average 1 metre, the depth about 30 cm. They are usually wok shaped.

6. The filling consists of fire-cracked stones and black earth coloured by charcoal.

7. Artefacts are rare.

There are three major directions in the attempts at interpreting the cooking pit sites (Thrane 1974). The first not very common one is that the sites are secular production sites (cf. Seeberg and Olesen 1971). Heidelk-Schacht and Fritz Horst saw the sites as expressions of fire cult and Horst suggested a relation to the fire cult known from the Mediterranean sphere and Central Europe (Horst 1976, 1985: 118ff, cf. Krämer 1966). According to this hypothesis the pits are simply fireplaces in which the fire was lit in order to be seen. A final major trend is to see the sites as meeting places where the function of the pits was the processing of food (Narmo 1996, Henriksen 1999, Gjerpe 2001). This interpretation is by no means a rejection of the possible cultic aspect of the sites, only of the pits.

It is a major problem for all attempts at interpreting the cooking pit sites that their constituent components, i.e. the cooking pits, are so anonymous. It is a mass material with little to offer except for charcoal and fire-cracked stones. Only under lucky circumstances are other finds like teeth, cremated bones, or pottery at hand. Therefore the interpreter is either left free to speculate or paralyzed by the absence of positive evidence. This situation has characterized the research so far.

Invisible structures? Cooking pits in Scania.

In order to get further in the understanding of the phenomenon it is necessary to make use of new excavation methods and to formulate questions in advance. This was done in connection with the Scanian West Coast Railroad project (1996-1999) during which 5 different sites (**Fig. 2**) with cooking pit concentrations were investigated (Olsson and Mattisson 1996, Martens 1999, Fendin 1999, Artursson 1999, Grønnegård 2000). Two of these sites were

cooking pit sites at Iron Age settlements (**Fig. 6**), three were specialized cooking pit sites. In all instances the same questions were asked and the same methodology was applied.

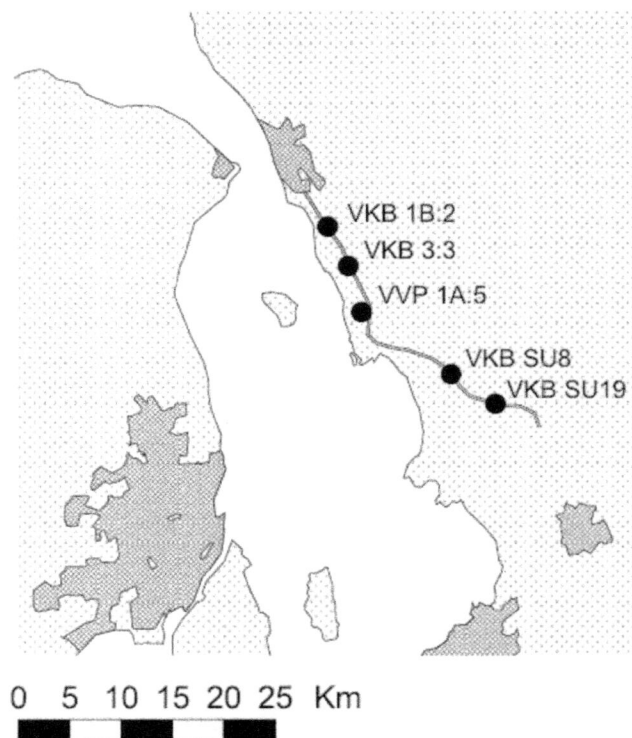

Fig. 2: Cooking pit sites investigated in connection with the Scanian West Coast Railroad project (1996-1998) (prep. by the author).

The most interesting location was the large specialized cooking pit site at Glumslöv backer (**Fig. 3**). The site which dates from the Bronze Age (Mont. II-V) (basing on 42 14C-datings) was situated at the hill side of the most marked hill at the east coast of Øresund. In advance it was therefore speculated that the site somehow corresponded with the Sund from which it would be visible, provided that the 2km wide strip of land in between was treeless. It was even suggested that the pits could have been signal fires or something similar for the seafarers coming into the Sund from Kattegat. During the planning of the excavations, these and other hypotheses were discussed and ways to try them out were figured out.

The first thing to try out was to see whether the cooking pit site was "isolated" or had relation to a nearby settlement. It turned out that the nearest contemporary houses were situated only 225m to the south and 20m higher in the landscape than the cooking pits (**Fig. 4**). The area between the houses and the cooking pits was occupied by traces after various settlement activities like pits and wells, leaving only a band of 50m width untouched. The settlement could be dated to Montelius

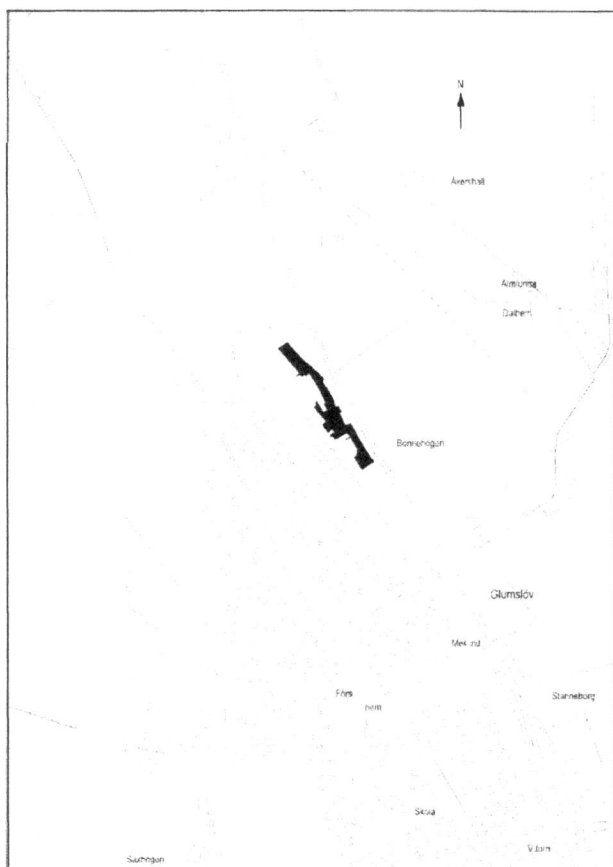

Fig. 3: The situation of the excavation at Glumslöv Backar, VKB 3:3, between the village Glumslöv and the high way E6, Scania (Fendin 1999, fig. 2.)

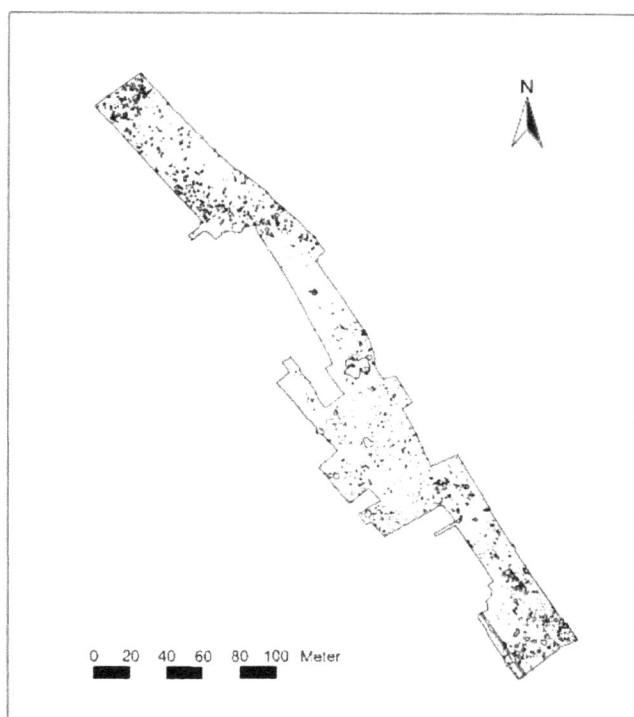

Fig. 4: Glumslöv Backar, VKB 3:3, Scania. The overall excavation plan. The house sites of the settlement were situated in the southernmost "pocket" of the excavation (prep. by the author).

IV-V on the basis of house typology and a large pottery material. Since the excavation was confined by the narrow limits of the coming railroad it was most likely only a fragment of the settlement that could be excavated. Therefore it cannot be excluded that the settlement had existed even at the start of the cooking pit site. While the house sites were located at a plateau at about 70m over sea level there were nearby burial mounds (Bonnehögen) even higher in the terrain (**Fig. 3**). They have not been subject to archaeological investigations but their location and size indicate that they were most likely built during the Early Bronze Age (Mont. I-III), thus indicating that even then the cooking pit site was not "isolated". Concerning Heidelk-Schacht's criteria about exposure, it is interesting to note that the settlement and the barrows occupy even more exposed locations than the cooking pit site. It is also important to note that the cooking pits in the northernmost part of the site are situated at the foot of the hill, and if the surrounding area was not completely deforested at the time they would not be "exposed" at all. In this connection it is interesting to note that Norwegian archaeologists have a diverging idea of the situation of the cooking pit sites since they suggest that they were located in sacred woods (cf. Farbregd 1986, Gustafson 1999).

The second question was the hypothesis of fire cult. The subsoil at the site was a heavy clay soil which was very sensitive to firing, and responded to exposure to high temperatures by red colouring. In spite of this it was immediately noticed that discolouring of the ground surrounding the cooking pits was very rare. An analysis of the soil in which the pits had been dug demonstrated that only one or two of the 323 unearthed cooking pits had been fired with temperatures above 300 degrees Celsius. This proved that the other pits had not been fired with open fire. However, far the majority of the cooking stones had been exposed to temperatures between 300 and 700 degrees Celsius, i.e. well above the threshold of red colouring of the local clay soil. It therefore had to be concluded that the pits were dug for using indirect heat from fired stones. If the site was a place for fire cult, then it was not practiced in the cooking pits, but even if there had not been any fire cult, there would have to be one or more fireplaces at the level surface for heating the cooking stones. The problem only was that the site at the time of investigation had been ploughed for more than a century and every physical trace of such a fireplace would by then have been erased.

A third question was whether there were traces of other activities at the site. There were almost no other types of features in the area of the cooking pits. This might have been due to the fact that the excavators had to go very deep into the ground in order to get a clean surface

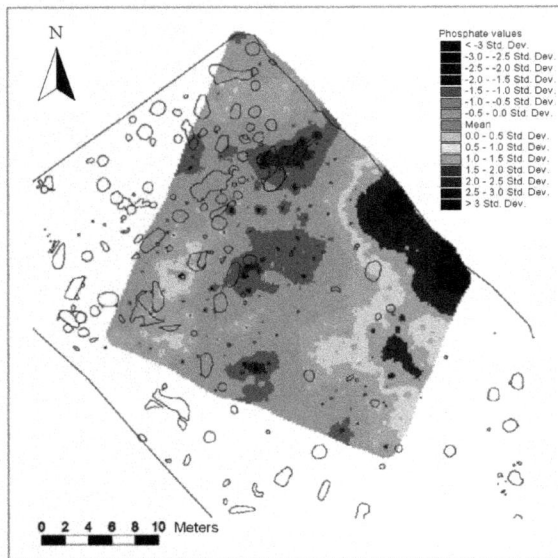

Fig. 5: Glumslöv Backar, VKB 3:3, Scania. Phosphate mapping of the northern end of the excavated area. "Gennemsnitsværdi" = average value. "Std. afvigelse" = Standard deviation (prep. by the author).

without modern disturbances, or it might be due to a matter of fact. A way to test this was by means of phosphate mapping. The cooking pits were organized as an arch-shaped band leaving an almost free area inside. Inside this area there were as mentioned only a few postholes and some "isolated" cooking pits. The post holes were carefully investigated and attempts were made at figuring out their purpose. After cleaning the excavation surface several extra times it was finally concluded that these postholes did not derive from prehistoric house constructions, but since many of them could be ordered in lines it was possible that they were remains of fences. An area in the northern end of the excavation which included both a part of the cooking pit arc, a part of the free inner area and traces of the possible fences was therefore chosen for mapping. Soil samples were taken within a grid with the side 1m. Samples used for the mapping were not taken in features and were checked for humus and other pollutions. The samples were analyzed by the German laboratory ABOLA by means of the calcinations method, a method coming very close to a total phosphate analysis (Bleck 1965, Zimmermann 2001). The result was surprising (**Fig. 5**). The highest phosphate values (red) were concentrated to the inner space. More variable values were found in the area among the cooking pits indicating that there might have been local fireplaces at the edge of the pits for heating the stones. However, the most interesting observation was that these two areas were separated by a band of lower phosphate values (blue), giving the impression that the two areas had been separated from each other by a barrier. In fact, this "blue" area coincided with two of the mentioned rows of postholes confirming the interpretation as traces of fences. In the northwest a "phosphate bridge" connected the central concentration

with the area of the cooking pits. This seems to indicate that on this spot there was an opening in the fence, and indeed there is an opening even between the cooking pits.

The interpretation of this pattern shall be dealt with soon. Before that a final analysis shall be mentioned. Also the soil below some of the cooking pits were phosphate mapped. This demonstrated that the pits in the northern half of the area had relatively high values, while the southern half was characterized by lower values, thus indicating a difference in use in different areas (**Fig. 6**). A most likely explanation to the difference in phosphate would be that the northern pits were used for cooking, the southern not. 14C-datings indicate that both areas were in use during most of the life time of the site. Thus the phosphate mapping indicates that the cooking pit site was organized in functionally specialized areas, an observation which was not traceable by other means.

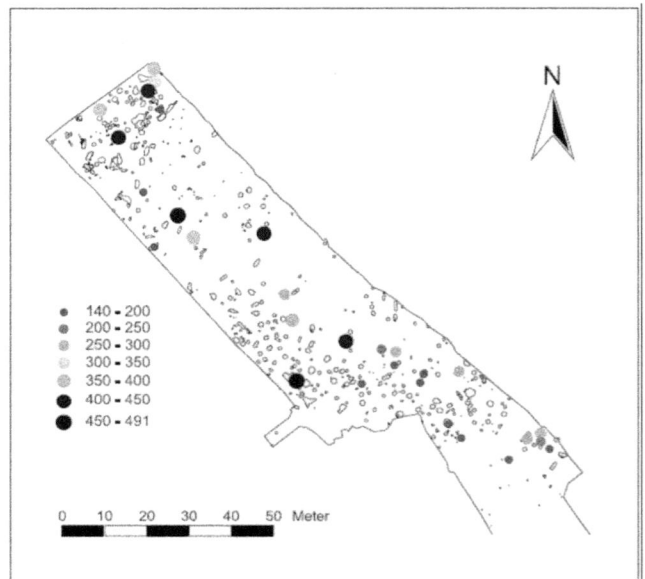

Fig. 6: Glumslöv Backar, VKB 3:3, Scania. Phosphate values (in ppm P) of the soil under selected cooking pits (prep. by the author).

In conclusion it may be said that the methods applied gave answers to the questions posed in advance, and so several of the hypotheses could be rejected. Apparently the assumption that the cooking pit site was more than just the cooking pits was the one assumption which stood ground. But the results also posed new questions. What happened inside the cooking pit "curtain"? High concentrations of phosphate may be produced by several factors, offal being one, cooking another. Maybe they had a huge fire here for heating the stones for the cooking pits, or maybe they ate their meals here, prepared in the northern cooking pits. In the latter case the phosphate would have been the trace of the resulting kitchen midden. A third possibility is that the concentration marks a place for depositing offal from sacrifices (animal or human). We have several indications that such cult places existed during the Late Bronze Age in Northern Europe, but normally traces of the sacrifices will only be preserved in wetlands. If the cooking pit site at Glumslöv was such a site the inner area would have been the sacred place surrounded by a fence and rows of cooking pits. In that case the main function of the pits could have been to contribute to the marking of the inner space as something sacred and secret by producing smoke and steam. That steam can function in a cultic context is known from ethnographic parallels (Runcis 1999). The place would then both be given an air of secret and mystery proper to its purpose. Though this is the bolder suggestion it is at the same time the most appealing.

Bibliography

ARTURSSON, M.
1999 Saxtorp. Boplatslämningar från tidigneolitikum-mellanneolitikum och romersk järnålder-folkvandringstid. Skåne, Saxtorp sn, Tågerup 1:1 och 1:3. Västkustbanan SU8, RAÄ 26. *UV Syd Rapport*: 79.

BLECK, R.-D.
1965 Zur Durchführung der Phosphatmethode. *Ausgrabungen und Funde*: 213-218

DEHNKE, R.
1968 Eine spätbronzezeitliche Kultanlage mit Feuerstellen in Bötersen, Kr. Rotenburg (Wümme). *Nachrichten aus Niedersachsen Urgeschichte Band 36*: 116-120.
1972 Kultfeuerstellen bei Narthauen im Kreise Verden. *Nachrichten aus Niedersachsen Urgeschichte Band 41*: 22-33.

FARBREGD, O.
1986 Hove i Åsen - kultstad og bygdesentrum. *Spor nr. 2.*

FENDIN, T.
1999 Boplats och härdgropsområde från bronsåldern vid Glumslöv. Skåne, Glumslöv sn, Övre Glumslöv 10:5, Västkustbanan 3:3. *UV Syd Rapport*: 39.

GJERPE, L. E.
2001 Kult, politikk, fyll, vold og kokegropfelt. *Primitive tider*: 5-17.

GRØNNEGAARD, T. J.
2000 Landsby i folkevandringstid. In Andersson, M., T. J. Grønnegaard, and M. Svensson, Mellanneolitisk pallisadinhägnad och folkvandringstida boplats. Skåne, Västra Karaby sn, Västra Karaby 28: 5, Dagstorp 17: 1, VKB SU19. *UV Syd Rapport 1999 (101)*: 25-40.

GUSTAFSON, L.
1999 En kokegrop er en kokegrop er en...? *Follominne 1999*: 6-13.

HEIDELK-SCHACHT, S.
1980 Bütow, Kreis Röbel - Siedlung oder Kultstätte der Jungbronzezeit? *Jahrbuch für Bodendenkmalpflege in Mecklenburg 1979*: 59-82.
1989 Jungbronzezeitliche kultfeuerplätze im Norden der DDR. In Schlette, I F. and D. Kaufmann (eds.): *Religion und Kult in ur- und frühgeschichtlichen Zeit*, Berlin, pp. 225-240.

HENRIKSEN, M. B.
1999 Bål i lange baner - om brugen af kogegruber i yngre bronzealder og ældre jernalder. *Fynske Minder*: 93-123.

HORST, F.
1976 Siedlung und Opferplatz der jüngeren Bronzezeit von Zedau, Ot. v. Osterburg (Altmark). *Zeitschrift für Archäologie, Bd. 10*: 121-130.
1985 *Zedau. Eine jungbronze- und eisenzeitliche Siedlung in der Altmark.* Schriften zur Ur- und Frühgeschichte Bd. 36, Berlin.

KRÄMER, W.
1966 Prähistorische Brandopferplätze. I Degen, R., Drack, W. & Wyss, R. (red.): *Helvetica Antiqua. Festschrift Emil Vogt.* Zürich: 111-122.

MARTENS, J.
1999 *Kogegrubeområde med grave og lertagningsgruber ved Säbyholm. Skåne, Säbyholm, Vej 1156, Plads 1A:5.* UV Syd Rapport (58) Lund: Arkeologisk undersökning.
2004 Kogegruber i Syd og Nord - samme sag? Består kogegrubefelter bare af kogegruber? In Gustafson, L., Heibreen, T. and Martens, J. (eds.): *Gåtefulle kokegroper. Internordisk seminar om kokegroper, Oslo 2001.* Varia.

NARMO, L. E.
1996 Kokekameraterne på Leikvin. Kult og kokegroper. *Viking 1996*: 79-100.

OLSON, T. and A. MATTISSON
1996 Plats 1B:2 - Härdområde och boplatslämningar från bronsålder. I Svensson, M. & Karsten, P. (red.): Skåne, Malmöhus län, Järnvägen Västkustbanan, delen Helsingborg-Kävlinge. Avsnittet Helsingborg-Landskrona (block 1-2). Arkeologisk förundersökning. *UV Syd Rapport 1996: 48*, Lund: 19-24.

RUNCIS, J.
1999 Reflektioner kring skärvstenshögar, mytologi og landskapsrum i Södermanland under bronsåldern. In Olausson, M. (ed.). *Spiralens öga. Tjugo artikler kring aktuell bronsåldersforskning.* Stockholm: RAÄ Arkölogiska undersökningar, Skrifter nr. 25: 129-158.

SEEBERG, P. and A. OLESEN
1971 Storudvinding af trækul. *MIV*: 48-51.

THRANE, H.
1974 Hundredvis af energikilder fra yngre broncealder. *Fynske Minder*: 96-114.

THÖRN, R.
1996 Rituella eldar. In: Engdahl, K. and Kaliff, A. (ed.): *Religion från stenålder til medeltid.*

Riksantikvarieämbetet. Stockholm: Arkeologiska undersökningar. Skrifter nr. 19: 135-148.

WEGEWITZ, W.
1943 Herdgruben in der Feldmark Tangendorf, Kr. Harburg, *Die Kunde 11*: 127-143.

ZIMMERMANN, W. H.
2001 Fosfatanalyse - et vigtigt bebyggelsesarkæologisk redskab. *Arkæologiske udgravninger i Danmark*: 21-43.

Ashes to Ashes: The Instrumental Use of Fire in West-Central European Early Iron Age Mortuary Ritual

Seth A. Schneider

> ...I reckon that its total length is the same as that of the Ister [Danube River]. The Ister rises in the land of the Celts, at the city of Pyrene (the Celts live beyond the Pillars of Heracles and are neighbours of the Cynesians who are the westernmost European people), and flows through the middle of Europe...
>
> (Herodotus II:33).

Introduction

Though Herodotus' geography might have been a little off when describing the location of the Danube River in the fifth century BC, the mention of the Celtic city of Pyrene near the headwaters of the Danube River has been linked to a well known early Iron Age (EIA) hillfort, known as the Heuneburg, where the use of fire and its destructive capabilities are well documented. The EIA Heuneburg hillfort is situated near the source of the Danube River in southwest Germany near the town of Hundersingen (**Fig. 1**). The hillfort had several pyrotechnic production areas of metals, glass, and jewellery during its early periods from 600 – 500 BC (Gersbach 1989; Kimmig 1983: 88). A mudbrick wall fortification enclosed the hillfort at this time as well, which was an uncommon building practice in west-central Europe during the early Iron Age (750 – 400 BC) and is analogous to similar structures found in the Mediterranean region.

The Heuneburg hillfort's mudbrick wall and Herodotus' mention of Pyrene, which has its root in the Greek word for fire, *pŷr*, suggest a correlation with the hillforts significant use of fire, especially in iron production. Herodotus wrote about Pyrene almost a century after its demise, but his information probably came from the writings of Hecataeus of Miletus, the "Father of Geography", who wrote about the Danube in the middle of the sixth century BC (Kurz 2005; Spindler 1983: 16).

The significance of pyrotechnic crafts in the Heuneburg hillfort's economic and social position in the EIA West Hallstatt zone (eastern France, northern Switzerland, and southwestern Germany) (**Fig. 2**) has been noted (Frankenstein and Rowlands 1978). Fire was also instrumental in mortuary contexts at this time as a means of defining social differentiation and in ritual activities that connected the living population with the ancestors. Cremation burials and charcoal concentrations, most likely ritual offering places, are evidence for the instrumental use fire in EIA mortuary contexts. This paper will focus on the relationship between funeral pyres, cremation burials and other pyro-activities in burial mounds during the EIA and the use of fire by elites as a means of maintaining their social, economic and religious power. Few researchers have made a direct connection between these pyro-cultural phenomena in cross-cultural mortuary contexts and evidence from recent excavations of burial mounds in the so called "Speckhau" mound group near the Heuneburg hillfort in southwest Germany (Arnold 1991a; Arnold 1995; Arnold and Murray 2002; Arnold et al. 2000, 2001, 2003; Kurz 2001: 38-40; Kurz and Schiek 2002: 38-40; Schneider 2003). The correlation between cremated remains and later offering activities taking place on existing mound surfaces that has been documented for these mounds deserves a more careful scrutiny.

Ethnographic ritual use of fire

The cultural role of fire is not just functional (heat and cooking) or technological (metal smelting/working, glass and ceramic production). Marcel Otte states that fire is "...[a] veritable agent of spiritual, mythical and social development" (Otte 2002: 7). The ritual uses of fire include 1) purification and 2) a way of transporting material to the Otherworld with cultural and social implications intertwined as individuals, groups, and objects pass through liminal states. Fire as a purification element frequently appears ethnographically to sanctify actions that have already occurred or are about to occur. For example, a recent re-interment of Native American human remains in the United States was preceded by a purification ceremony of fire within the burial pit to cleanse the area and prepare the earth to accept the offering (Kuhn 1987). Prehistoric examples of such practices also exist. Ash and charcoal concentrations found above pre-contact burials in the Gentleman Farm Mound, a Langford Tradition (AD 1200 – 1500) burial

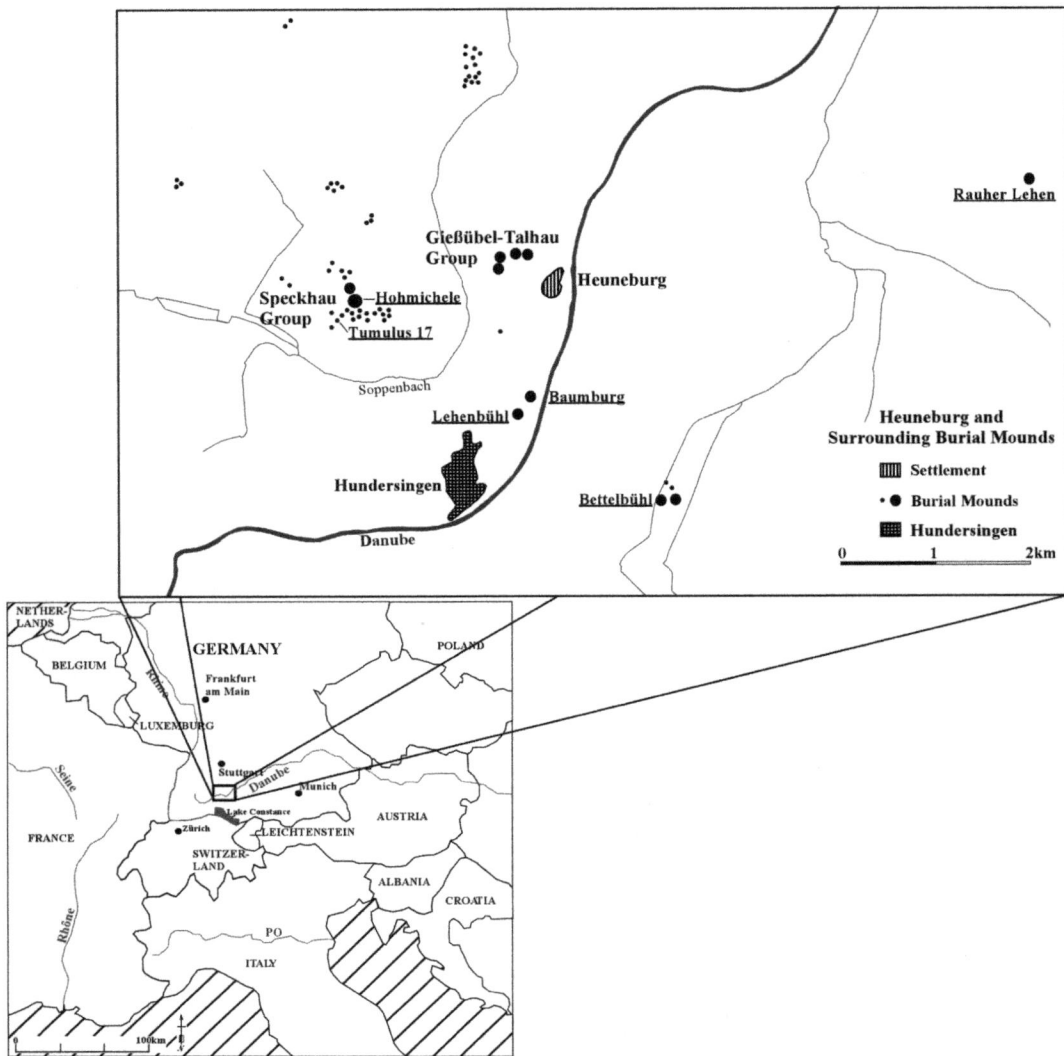

Fig. 1: Map showing the location of the Heuneburg hillfort and the town of Hundersingen (after Arnold 1991a and Reim 2002).

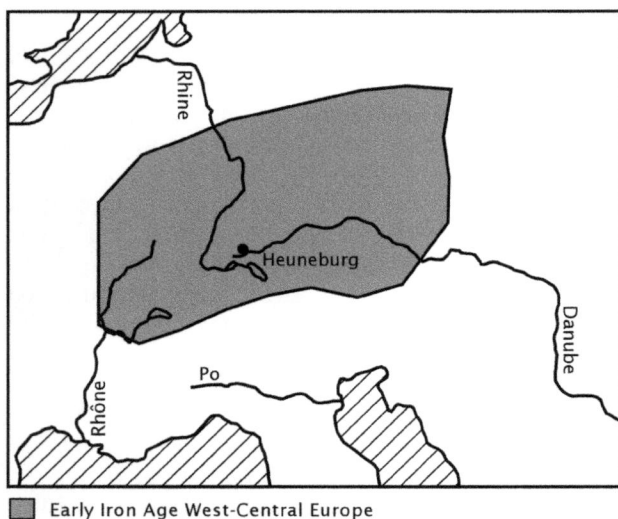

Early Iron Age West-Central Europe

Fig. 2: Map showing the extent of the West Hallstatt zone during the early Iron Age (after Arnold 1991a, 1995).

mound in the Great Lakes region of the United States, have been interpreted as mortuary "watch fires" similar to those used by historic Native American tribes in their mortuary practices. James Brown and others argue that "the mortuary fires of this historic period were usually kept burning over the grave for four days and four nights" (Brown *et al.* 1967: 4). Incense and smoke in many religions today are used to signify the transmission of thoughts or prayers to the heavens, such as incense burners used by the Catholic Church during special ceremonies at Easter and Christmas.

Transmissions to the Otherworld by fire can be in the form of cremations and physical offerings as well. Fire used to cremate the deceased is a tool that allows the soul of the individual to be transported into the afterlife. Sergei Kan's (1989) study of the Tlingit potlatch ceremony in Northwest Coast North America describes cremation as necessary to get the soul across the celestial threshold because full members of the tribe were not strong enough spiritually to do so without the aid of fire. Slaves, shamans, and great chiefs were treated differently in Tlingit society. Slaves were left on the beach margin or in the shallow waters when they died, watery environments. Shamans and great chiefs, on the other hand, were the only individuals inhumed because they were thought to have the power to transmit their souls into the Otherworld without assistance (Kan 1989: 120-121). The opposite can be true as well, where fire is a necessary tool in assisting individuals to crossover into the afterlife.

The act of transmission by fire to the Otherworld is an event that not only provides a means of disposing of the body and soul of the individual but is also a rite of passage for the community that is etched into the minds of the participants (Geertz 1980: 32; Lungu 2002: 140; Van Gennep 1960). Vasilica Lungu argues that the portrayal of corpses being consumed by the flames on ancient Greek pottery and in poetry, for example in Homer's epic the *Iliad*, "transformed the memory of the dead in a tangible and credible legend, i.e. history…The fire had the role to transform the human's bloody flesh into exemplary memory of the soul" (Lungu 2002: 140). For example, during the Negara period in Balinese history cremation was one medium used to display elite status, with larger funeral pyres being processed through the villages as it makes its way to the spot of cremation and swarms of people participated in the event (Geertz 1980: 117). Geertz comments that even though cremation was practiced by all members of Balinese society, "…cremation (*ngabèn*) was, in fact, the quintessential royal ceremony. Not only was it the most dramatic, splendid, sizable, and expensive; it was the most thoroughly dedicated to the aggressive assertion of status" (Geertz 1980: 117).

Ian Morris (1987: 46) argues that in Homeric Greece the funeral pyre and burial mound was a sign of elite status. The energy expenditure that went into gathering wood for the pyre for Hector's funeral ceremonies (nine days) as compared to the everyday soldiers who fell on the battle field (one day), signified Hector's place in society. Energy expenditure was also expressed in the size of the burial mound placed over the individual. The size of the burial mound in Homeric Greece was also directly proportional to an individual's status, so the higher the status the larger the mound. Morris makes the comment that the whole Achaean host would assist in gathering wood and raising the burial mound over someone like Agamemnon, while for lesser individuals a more modest funeral would have been in order. The physical manifestation of status through the burial mound allowed the descendants to share in the individual's status (Morris 1987: 46). Homeric Greek mortuary practices have been referred to when interpreting similar practices in the early Iron Age of west-central Europe (Kurz 1997; Schiek 1981; Zürn 1964). However, the instrumental use of fire by high status individuals is not a synchronic event but maybe continued by subsequent generations in the form

of pyrotechnic activities on existing mound surfaces (Keller 1871; Schneider 2003).

Fire in the Early Iron Age

Fire played a significant role in the ritual behavior of the late Bronze Age and early Iron Age (1200-400 BC) in the west Hallstatt zone of west-central Europe. Charcoal features (*Feuerstellen*), charcoal concentrations (*Holzkohlennester*) in mounds, fire offering places (*Opferfeuerplätze*), cremation areas (*Scheiterhaufenplätze*), cult sites (*Kultstätten*) and other burnt cultural remains found in and around mortuary sites all attest to the ritual significance of pyrotechnic activities. Traditionally, archaeologists have not focused much attention on such features in mortuary contexts because of the difficulty in making connections between the actual disposal of the body and related non-burial activities. The focus on cremation burials as burial rites during both time periods suggests the importance of these pyro-cultural phenomena.

The ritual use of fire during Hallstatt C/D attests to the importance of pyro-instruments like funeral pyres for cremation and areas of burnt offerings (*Brandopferplätze*) in cult and mortuary activities in west-central Europe (Dietrich 1998; Kurz 1997; Reim 1988). Areas of burnt offerings containing large quantities of charcoal, burned animal bones, and secondarily burned pottery sherds found in association with mortuary remains or in isolated areas have been identified at sites in the Alps and north of the Alps dating from the late Bronze Age to the early Iron Age. These offering places (*Opferplätze*) are mainly situated on hills or mounds in the form of large deposits of pyrotechnic materials. Such deposits have been compared to similar Bronze Age features called ash altars at sites like Olympia and on the island of Samos in Greece that were used for religious activities, including making offerings to the gods (Krämer 1966). A subset of the Opferplätze, called *Opferrinnen* (offering trenches), are known to be associated with Iron Age Greek burials. These usually contained ceramic vessels and in some instances pyre remains were deposited as part of the offering (Morris 1987; Snodgrass 1971). The deposits in the Opferrinnen do not appear to have been items curated from the original funeral pyre, however. Similar offering places located in the Alps and just north of the Alps may be due to influence from the Mediterranean region, though whether their function was the same as that of the Greek examples is unknown.

Burnt offering areas and places where cremations took place are known in and around early Iron Age tumuli in west-central Europe (Dietrich 1998; Kurz 1997; Reim 1988). For example, the Iron Age cemetery at Rottenburg on the Neckar River in Baden-Württemberg contained burials from Ha C to La Tène C (750-150 BC), with 56 burial mounds that date to the Late Bronze Age (Ha C) and early Iron Age. Fifty of these burial mounds contained central cremation burials and, in some cases, they were erected over the Scheiterhaufenplatz. Seventy-nine cremation burial pits dating to the Ha C/D periods were found in and around burial mounds at Rottenburg, further evidence of the intense pyro-activity at the site (Reim 1988). The pyrotechnic features seen at Rottenburg may be a continuation of similar activity during the Urnfield Culture of the late Bronze Age when the dominant mortuary practice involved placing cremations in urns in flat cemeteries (Murray 1992: 98). The recovery of secondarily burned sherds and pyre remnants in various mound contexts from the Hallstatt mound group at Heidenheim-Schnaitheim also supports the idea that the use of pyrotechnics in mortuary ritual was a wide spread phenomenon in Baden-Württemberg and throughout the West Hallstatt zone (Dietrich 1998: 35).

The funeral pyre as a status symbol

The early Iron Age of west-central Europe is characterized as a society with a complex social organization, (Arnold 1991a, 1996, 1999; Bintliff 1984; Bittel *et al.* 1981; Burmeister 2000; Dietler 1995, 1996; Eggert 1988; Fischer 1982; Rieckoff and Biel 2001) and information about this complex social organization comes primarily from the differential treatment of individuals in burial contexts. Most EIA burials come from burial mound contexts, though in a few cases they are found in flat burials, such as the burial inside the walls of the Heuneburg hillfort on the Danube River (Dämmer 1974) or the Ha D burials located between tumuli at Rottenburg on the Neckar River (Reim 1988) and Heidenheim-Schnaitheim (Dietrich 1998). An individual's social importance has been interpreted as having been based on where they are located in the burial mound and the type, quantity, and quality of grave goods associated with them. Bettina Arnold (Arnold 1991b, 1996, 2001) has argued that the assemblage of grave goods for paramount elites was a 'vocabulary' that separated these individuals from the rest of society. The elements of mortuary ritual for a paramount elite burial consisted of: 1) gold neck ring; 2) wheeled vehicle; 3) drinking/feasting equipment of bronze or gold; and 4) Mediterranean imports (Arnold 2001: 216). The central chamber burials in the Hochdorf tumulus near Stuttgart in Germany (Biel 1985) and in the Vix tumulus on the Seine River in France (Joffroy 1962; Megaw 1966) are two examples of early Iron Age paramount elite burials circa 500 BC. The central chambers of these two burial mounds were intact, but in most cases the central chambers of early Iron Age tumuli were looted in antiquity or disturbed during the early explorations of burial mounds.

Secondary burials that come after the central chamber interment can be of equal or lesser social standing based on the material remains found with them, as well as adding new layers to the mound (Arnold 1991a: 65-66). Ian Morris (1987: 46) shows that in Homeric Greece the size of a mound was a reflection of the individual's social standing. The expenditure of time and energy on constructing the burial mound and its continued use as a mausoleum for the deceased relate to the importance of the individual in the central chamber in the West Hallstatt zone. The Hohmichele (13 m high with a diameter of 80 m) (Riek 1962) and the Magdalenenberg (6.5 m high with a diameter of 102 m) (Spindler 1983: 134), the two largest burial mounds of the Iron Age in west-central Europe, would have required considerable amounts of time and energy to construct, for example, and while both central chambers were looted, they can be assumed to have been commensurately rich.

Energy expenditure did not just go into the physical construction of the burial mounds, but also into the preparation of the body for both inhumation and cremation burials (Arnold 1991a: 70-73, 1995: 43). Wood, a necessary commodity in the early Iron Age for use in domestic, pyrotechnic (metal, glass, and ceramic production), and structural purposes, was used for both inhumation and cremation burials. A caveat to the use of wood in cremations is that another form of combustible material, dung from cattle, would have been available in the EIA (Rieckoff and Biel 2001:142-143). The area covered by an EIA funeral pyre averaged around 3 m^2 (Kurz 1997: 70). For example, the central chamber of Tumulus 18 in the Speckhau mound group had two log-like depressions set into the floor of the area covered with large quantities of charcoal at the base and in the center of the mound, which is evidence that they formed part of the base of the funeral pyre (Arnold *et al.* 2003). The depressions are approximately 2.5 m in length and 3 m wide, which represents an area of 7.5 m^2 covered by the pyre (**Fig. 3**). The distribution zone of the charcoal remains from the pyre is much greater, 4.8 m x 4.3 m, or 20.6 m^2. The central chamber of Tumulus 18 is slightly smaller than other cremation burials located in the Hohmichele, with Grave IX having an area of 20.8 m^2 (Kurz 1997: 71, Tab. 5), while the area of central chamber cremation in Tumulus 17, which is approximately 25 m^2 (Table 1). Wood used to construct funeral pyres for cremation of the deceased would have added to the expenditure of this commodity. As Arnold argues "the procurement of wood for a funeral pyre therefore must be factored into the overall social cost of a cremation..." (Arnold 1991a: 72). Morris' (1987: 46) observation about the size of funeral pyres in Homeric Greece was probably true in the early Iron Age context as well: the larger the funeral pyre, the higher the social standing of the individual.

The Speckhau Mound Group

Exploration of the early Iron Age burial mounds in the Speckhau group associated with the Heuneburg hillfort on the Danube River in southwest Germany has produced evidence of various kinds of pyro-activity (**Fig. 4**).

Table 1. Size of cremation burials from systematically excavated tumuli in the Speckhau mound group.

Tumulus	Grave	Burial Type	Size	Area
Hohmichele	Grave IX	Cremation	4 m x 5.2 m	20.8 m^2
Hohmichele	Grave X	Cremation	4 m x 2 m	8 m^2
Hohmichele	Grave XI	Cremation	3.25 m x 4.5 m	14.63 m^2
Hohmichele	Grave XII	Cremation	3 m x 3 m	9 m^2
Hohmichele	Grave XIII	Cremation	2 m x 3.5 m	7 m^2
Tumulus 17	Central Chamber	Cremation/Inhumation*	5 m x 5 m	25 m^2
Tumulus 18	Central Chamber Area	Cremation	4. 8 m x 4.3 m	20.6 m^2
Tumulus 18	Grave 2	Cremation**	.85 m x .4 m	0.34 m^2

*The central chamber of Tumulus 17 is a bi-ritual burial containing the cremated remains of a probable male individual and the inhumation of a probable female individual, based on grave goods (Arnold *et al.* 2001: 69).
** Grave 2 of Tumulus 18 has been designated, preliminarily, as a cremation burial based on the concentration of cremated remains, a comprehensive report and interpretation of Tumulus 18 and 17 is forthcoming (Bettina Arnold, personal communication).

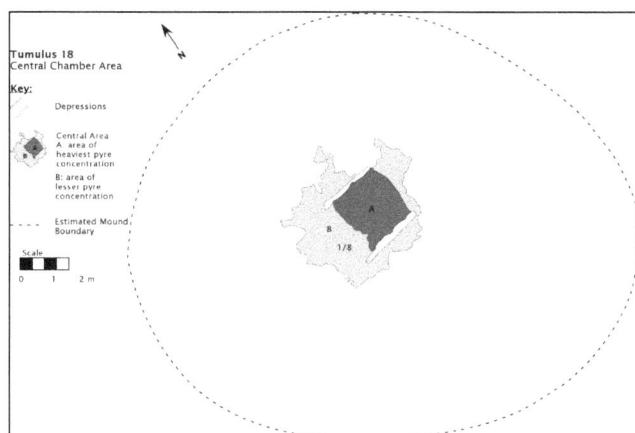

Fig. 3: Plan view of central area in Tumulus 18 showing the location of the two log-like depressions (after Arnold *et al.* 2003).

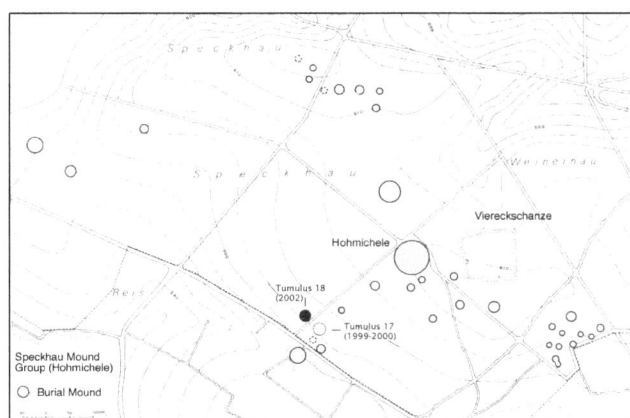

Fig. 4: Map of the Speckhau mound group with systematically excavated mounds (the Hohmichele, Tumulus 17, and Tumulus 18) indicated (illustration: Kurz and Schiek 2002: ill.1).

Gustav Riek's (1962) systematic excavation of the Hohmichele in the 1930s uncovered thirteen burials, five of which were cremations (Graves IX-XIII).

The Hohmichele yielded evidence of non-burial pyrotechnic activity as well, with 23 charcoal features found throughout the mound (including the four charcoal packets near inhumation Graves II-V) and fire offering places (**Fig. 5**). The fire offering places differ from the other charcoal features in the Hohmichele because stone slabs were found in association with the charcoal. Based on profile drawings, most of the nineteen charcoal features found in the mound fill were at stratum transitions and at angles following the slope of the mound surface. This suggests that these features were placed on the existing surface of the mound before they were covered over with mound fill. Wooden structures with rounded post supports, *Rundholzanlagen*, most likely platforms, were also identified at strata transitions. The Gießübel-Talhau mounds near the Heuneburg hillfort dating to Ha D2 and D3 (530-400 BC) were erected later than the Speckhau mounds, whose central burials date to Ha D1 or earlier, and show no evidence of such pyro-ritual activity. This suggests either a diachronic shift in burial mound mortuary practices, like the shift from ceramic to metal vessels seen from Ha D1 to Ha D2/D3 also seen in the Heuneburg mounds (Kurz and Schiek 2002; Rieckoff and Biel 2001: 177; Riek 1962; Schneider 2003: 23-24), or a functional/ideological distinction between the two groups.

The role of fire in association with funerary activities in burial mounds appears to decline during Ha D2 and D3 with the shift to inhumation as the dominant burial rite (Murray 1992; Rieckoff and Biel 2001). However, charcoal features in the outer mantle of Tumulus 17 in the Speckhau mound group suggest that fire continued to play a role in the funerary activities of late Hallstatt people (Schneider 2003). The pyrotechnic features are similar to those found in the Hohmichele and appear at or near boundaries between strata. Artifact seriation and

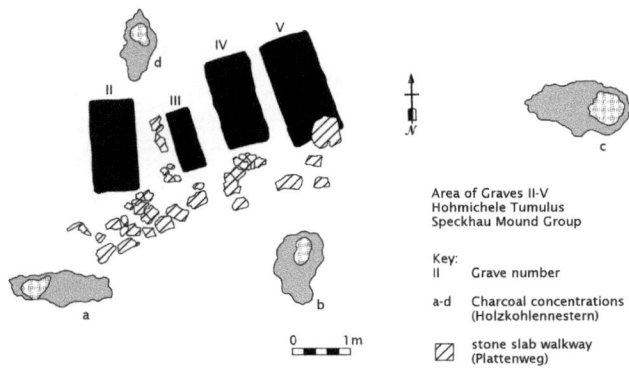

Fig. 5: Planview showing area around Graves II-V in the Hohmichele. The four shaded amorphous features around the graves are Holzkohlennester (charcoal concentrations) (after Kurz and Schiek 2002 and Riek 1969: 55).

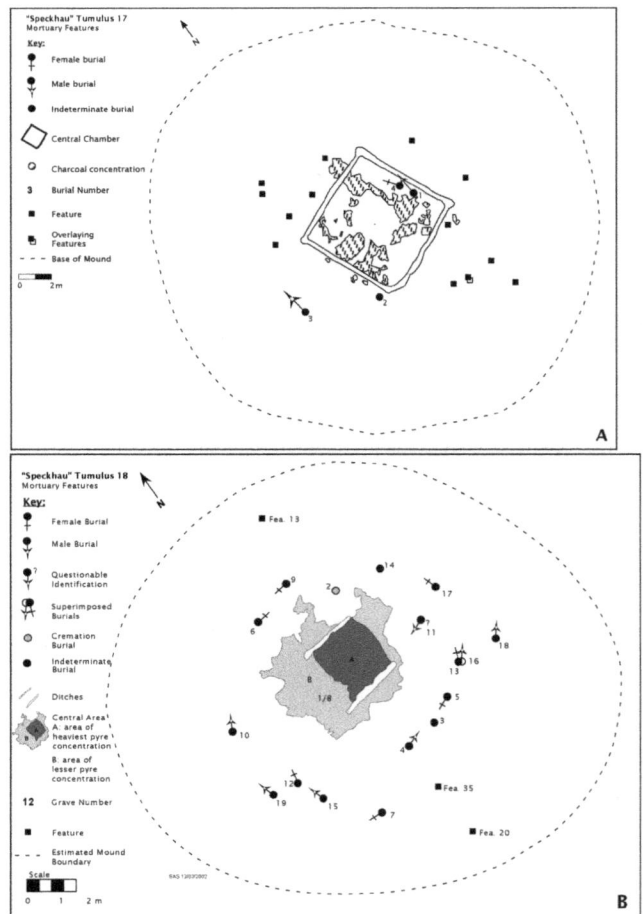

Fig. 6: Plans of Tumulus 17 (A) and Tumulus 18 (B) showing the location of the central chambers and secondary burials (after Arnold *et al.* 2003).

radiocarbon dates from burials and features indicate a use life of 150 years (600-450 BC) for the tumulus (Arnold *et al.* 2000, 2001). Some remains from the cremation burial in the central chamber were deposited outside the central cremation grave during the construction of the mound core, but other materials from this cremation were curated and deposited in later pyrotechnic features in the outer mantle (Schneider 2003). The combination of dates and placement of features within the mound reflect a diachronic use of fire in funeral activities at Tumulus 17 consistent with use of the mound as the focus of fire-based ancestor veneration even after the dominant mortuary ritual had shifted to inhumation. This paper places the Tumulus 17 features in the context of cross-cultural examples of such phenomena, and suggests possible interpretations of the uses of fire in Iron Age mortuary ritual.

Evidence from Tumulus 17 and 18 indicates that mortuary practices continued in the Speckhau mound group even during Ha D2/D3 (Arnold and Murray 2002; Arnold *et al.* 2000, 2001, 2003). Excavation of these burial mounds between 1999 and 2002 by the "Landscape of Ancestors" project uncovered 21 secondary burials in the two mounds (**Fig. 6**).

At least five of these could be dated to Ha D2 and D3, contradicting the traditional view that the Speckhau group was abandoned after the Ha D1 destruction of the mudbrick wall Heuneburg, with mortuary activity occurring only in the immediate vicinity of the Heuneburg hillfort (Gersbach 1969). The central chamber of Tumulus 17 was a bi-ritual interment containing the inhumation of a probable female and a disturbed cremated male primary interment. The central chamber in Tumulus 18 appears to represent the remains of an *in situ* funeral pyre, whereas the absence of reddened, fire-hardened earth and the thinner charcoal layer of the Tumulus 17 central chamber suggests the body in this burial was cremated elsewhere and pyre remains were then transported to the spot. The tumulus 17 central

chamber dates to approximately 600 BC based on radiocarbon dates (Arnold *et al.* 2001: 69), while Tumulus 18 may date to as early as 700 BC (Arnold *et al.* 2003).

Curation and charcoal features in Tumulus 17

Thirty-one non-burial features were identified in Tumulus 17. These primarily involved pyrotechnic or offering activities similar to those documented in the Hohmichele (Schneider 2003). Charcoal concentrations constituted the largest group of pyrotechnic features, while burnt earth and secondarily burned ceramics associated with charcoal were also found. Feature 8 was situated near the outer edge of the mound in the mound mantle near the boundary of Stratum 3 and 4 (**Fig. 7**).

The feature was a concentration of pottery with charcoal and burnt earth on top of a layer of pebbles. Stratum 5/Feature 10 was a significant feature within the mound core that became part of the mound stratigraphy between Stratum 9 and 8 when pyre remains containing large

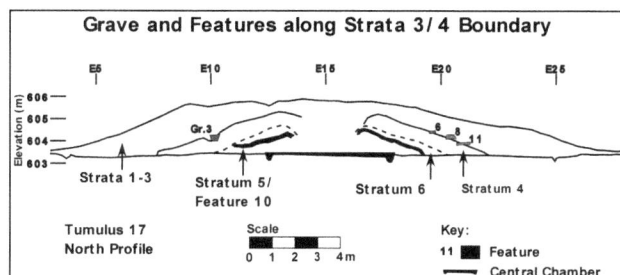

Fig. 7: Idealized mound profile of Tumulus 17 showing location of Grave 3 and features along the boundary between Strata 3 and 4.

quantities of charcoal and pottery associated with the cremated individual in the central chamber were deposited on the existing mound surface. An association between the central chamber and Stratum 5/Feature 10 was confirmed by the refitting of pottery sherds from Feature 46 within the central enclosure to sherds in Stratum 5/Feature 10. Refits between features in the mound mantle and core were also identified and come from cultural features that were separated by a distance of 8 m. Rim fragment refits from Stratum 5/Feature 10 and Feature 8 suggest the curation of materials (**Fig. 8**).

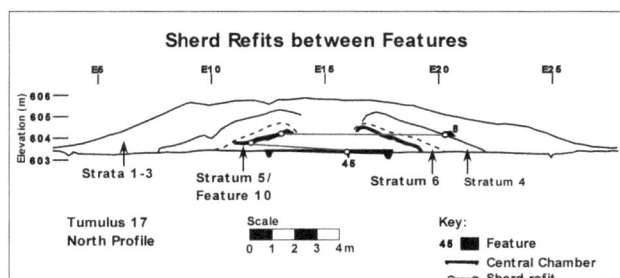

Fig. 8: Idealized mound profile of Tumulus 17 showing location of refits of sherds between features (after Schneider 2003: 168).

Many of these fire-related features in Tumulus 17, as in the Hohmichele, were found at or near the boundaries between strata, suggesting that internal subdivisions of late Hallstatt mounds had some ideological significance. Grave 3, situated at the boundary between Stratum 3 and 4, is significantly later in date than the central chamber, which was deposited around 600 BC. Based on fibulae in the Grave 3 assemblage, the burial can be dated to around 450 BC. If Feature 8 was coeval with Grave 3, as its position suggests, then 150 years had passed between the use of this ceramic vessel as part of the funeral assemblage of the cremated individual in the central chamber and the ritual activity that produced Feature 8 and led to the deposition of the refitting sherd. Given the features' locations, the paths of refit sherds from the central chamber, Stratum 5/Feature 10, and Feature 8, suggest that time elapsed between each depositional episode and the ritual activity that occurred later on the

successive mound surfaces. Sherd refits from different locations in the Hohmichele link graves and pyrotechnic non-burial features, additional evidence for curation of remains from earlier cremations (Kurz 2001; 2002) (**Fig. 9**).

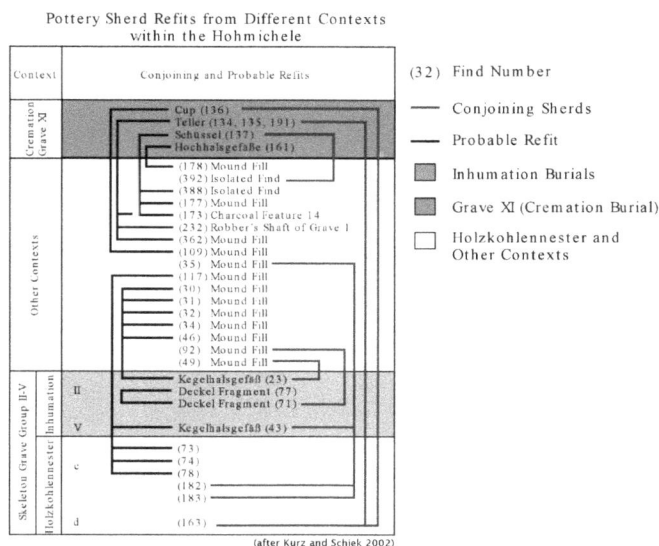

Fig. 9: Chart showing refits and probable refits of sherds between different locations in the Hohmichele. Lines on the left represent probable refits and lines on the right represent refits (after Kurz and Schiek 2002).

The pyro-ritual activity documented in the Speckhau mound group may represent purification acts at these mounds before the deposition of the next layer of mound fill. A similar complex of features is known from Switzerland. In 1871 Keller interpreted six to seven charcoal features and a circle of stones on a pre-existing mound surface in the Zollikon burial mound near Zürich as areas where fires were lit to represent the first stage ending the burial ceremony of the central chamber interment (**Fig. 10**). He suggested that these features represented a capping/ending ritual for at least the central portion of the mound. They may also represent purification with fire before the next layer was added to the mound. The central chamber of the Zollikon tumulus contained a number of ceramic vessels and fragments of these vessels could have been added to the ritual offerings, which might link these features to the central chamber. No ceramic material was mentioned in association with the charcoal features, and such an analysis was not carried out at the time (Keller 1871).

Early excavations like those at Zollikon can assist in determining the geographic range of pyrotechnic and capping rituals in other regions of the West Hallstatt zone during the late Hallstatt period. Re-excavation of unsystematically investigated late Hallstatt mounds with similar non-burial features, and the analysis of material from such excavations, represent avenues of future research. For example, the antiquarian A. Witscher's

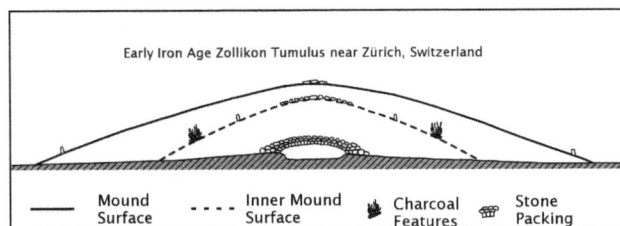

Fig. 10: Idealized profile of the late Hallstatt Zollikon Tumulus near Zürich, Switzerland. The inner mound surface has charcoal features represented as fires and upright stones that are described as forming a ring pattern (after Keller 1871).

description of numerous charcoal features in Tumulus 4 of the Roßhau mound group in the Heuneburg area suggests pyrotechnic ritual activity occurred in other Heuneburg mound groups as well (Kurz and Schiek 2002: 133-134) (**Fig. 11**). Excavation of a mound in this cluster could prove illuminating.

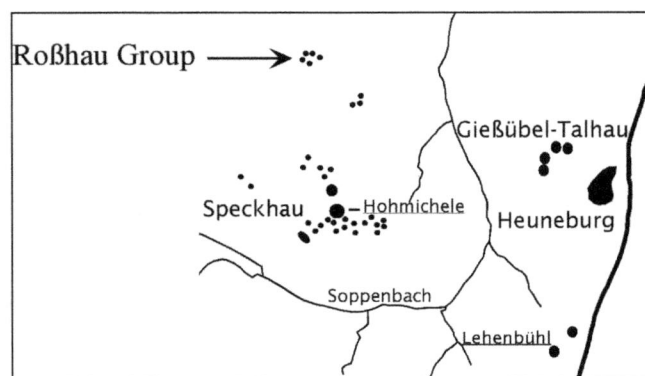

Fig. 11: Map showing location of Roßhau mound group in relationship to other early Iron Age features in the Heuneburg region (after Arnold 1991a).

Ancestor veneration and Tumulus 17

The features in Tumulus 17 clearly represent activities occurring before, during or after the placing of burials in the mound, but several questions may be posed based on this observation. When can a burial mound be said to have ceased to be used in a mortuary capacity? This is one of the questions in burial mound mortuary studies that is difficult to answer, since mounds may see later use in the same capacity by different people, or by the same people practicing a different tradition (Hingley 1996; Holtorf 1998). Some of the features in Tumulus 17 may represent rituals associated with the capping of the mound, indicating that it would no longer be used as a resting place for the dead. Another possibility is that the addition of each new layer or burial placed on or in the mound was marked, for example, by purification ceremonies involving the building of small fires. Some of these features may represent altars used as a way of

making contact with the spirit world, in order to keep some cosmological balance, or as offerings to the individuals or dead ancestors who are seen as residing in the confines of the mound.

The non-burial features in Tumulus 17 and their locations suggest that the ritual activity that produced them had earlier burials as a point of reference, indicating that these were planned or structured deposits and not random occurrences. It is likely that most non-burial features in the mounds are somehow related to one or more burials in the individual tumuli. The link between burials and non-burial features represents the burial mound's diachronic use-life, demonstrating that the Speckhau mound group was the focus of ritual activity much longer than previously thought. This also has implications for our understanding of the social organization of early Iron Age people in the Heuneburg area.

The curation of materials from the pyre of the cremated individual in the central chamber of Tumulus 17 supports the idea that late Hallstatt people conducted a form of ancestor veneration in their mortuary activities, with the individual in the central grave represented *pars pro toto*, meaning a portion was used to represent the whole. Tumulus 17 as a mortuary structure can be seen as analogous to the "ancestral hall," either familial or lineage, containing the remnants of the ancestor and selected descendants, reaffirming their association by interring curated burial remains in later ritual activity at the mound. Another possibility, based on the highly structured social system of the early Iron Age, was that ancestor veneration conducted at Tumulus 17 was a means for ruling elites to legitimize power over land and people.

Ancestor veneration has been identified in the early Iron Age in the form of anthropomorphic *stelae* placed on the summits of burial mounds (Bittel 1981: 90-92; Raßhofer 1998) (**Fig. 12**). The *stelae* have been interpreted as representations of avatars or lineage founders, but none have been found in association with tumuli in the Heuneburg area. Ancestor veneration in the Heuneburg area did not necessarily manifest itself in the form of mound markers alone, but as part of pyrotechnic rituals performed on existing surfaces of the burial mounds in the Speckhau group. The lack of pyrotechnic features in the Gießübel-Talhau mounds and the continued ritual activity in the Speckhau group suggests that the ruling elites of the Ha D2/D3 Heuneburg used ancestor veneration as a means of sanctifying their secular and ideological power. The Speckhau group may have remained the "ancestral mortuary seat" even after some elites began new mounds closer to the hillfort.

The use of fire in mortuary ritual remained an important aspect of elite signalling in the Heuneburg area even though a shift in burial practices from cremation to inhumation took place from Ha D1 to Ha D2/D3. The burial mounds represent the upper echelon of the early

Fig. 12: Hirschlanden stela with conical birch bark hat, neck ring, and dagger, which is ver similar to what was found in the central chamber of the Hochdorf tumulus (after Spindler 1983).

Iron Age and the energy expenditure required to both cremate and inter the individuals under a burial mound with elaborate grave assemblages in the central chambers of Tumulus 17 and 18 signifies their prestige in early Iron Age society. Later Heuneburg elites reaffirmed their position in society by recreating the mortuary ceremony of their ancestor with fire and a curated piece from the original funeral pyre which linked them directly with the past. Evidence for such activities has been identified in the Hohmichele and Tumulus 17 and the analysis of the Tumulus 18 features and burials and excavation of other mounds in the vicinity surrounding the Heuneburg, such as Tumulus 4 in the Roßhau group, may provide additional evidence for the hypothesis that fire, curation and ancestor veneration were an instrumental part of early Iron Age society in this region.

Acknowledgements

I would like to thank Bettina Arnold and Christine Hamlin for comments on earlier drafts of this paper. Any omissions or errors are my own. Thanks also go to the National Geographic Society and the States Monuments Office (Landesdenkmalamt) of Baden-Württemberg, Germany for their financial support of the "Landscapes of Ancestors" project. Thanks to Dragos Gheorghiu for arranging the session "Fire as an Instrument: The Archaeology of Pyrotechnologies" at the 9th Annual Meeting of the European Association of Archaeologists in St. Petersburg, 10 – 14 September, 2003. A special thank to Petra Nordin for her work in allowing my paper to be presented in the session.

References

ARNOLD, B.
    1991a *The Material Culture of Social Structure: Rank and Status in early Iron Age Europe.* Unpublished Ph.D. thesis, Harvard University. UMI, Ann Arbor.
    1991b The Desposed Princess of Vix: The Need for an Engendered European Prehistory. In Walde, D. and N. Willows (eds.) *The Archaeology of Gender. Proceedings of the 22nd Annual Chacmool Conference*, Archaeological Association of the University of Calgary, Calgary. pp. 366-374.
    1995 The Material Culture of Social Structure: Rank and Status in early Iron Age Europe. In Arnold, B.and D. Blair Gibson (eds.) *Celtic Chiefdom, Celtic State*, New Directions in Archaeology. Cambridge: Cambridge University Press. pp. 43-52.
    1996 'Honorary Males' or Women of Substance? Gender, Status, and Power in Iron-Age Europe. *Journal of European Archaeology* 3(2): 153-168.
    1999 'Drinking the Feast': Alcohol and the Legitimation of Power in Celtic Europe. *Cambridge Archaeological Journal* 9(1): 71-93.
    2001 The Limits of Agency in the Analysis of Elite Iron Age Burials. *Journal of Social Archaeology* 1(2): 210-224.
ARNOLD, B. and M. L. MURRAY
    2002 A Landscape of Ancestors in Southwest Germany. *Antiquity* 76: 321-2.
ARNOLD, B., M. L. MURRAY and S. A. SCHNEIDER
    2000 Untersuchen an einem hallstattzeitlichen Grabhügel der Hohmichele-Gruppe in "Speckhau", Markung Heiligkreuztal, Gde. Altheim, Kreis Biberach. *Archäologische Ausgrabungen in Baden-Württemberg 1999*: 64-68.
    2001 Abschließende Untersuchungen an einem hallstattzeitlichen Grabhügel der Hohmichele-Gruppe im "Speckhau", Markung Heiligkreuztal, Gde. Altheim, Kreis Biberach. *Archäologische Ausgrabungen in Baden-Württemberg 2000*: 67-70.
    2003 Untersuchungen an einem zweiten hallstattzeitlichen Grabhügel der Hohmichele-Gruppe im "Speckhau", Markung Heiligkreuztal, Gde. Altheim, Kreis Biberach. *Archäologische Ausgrabungen in Baden-Württemberg 2002*: 80-83.
BIEL, J.
    1985 *Der Keltenfürst von Hochdorf.* Konrad Theiss Verlag, Stuttgart.

BINTLIFF, J.
1984 Iron Age Europe, In the Context of Social Evolution from the Bronze Age Through to Historic Times. In Bintliff, J. (ed.) *European Social Evolution: Archaeological Perspectives*, Bradford: University of Bradford. pp. 157-226.

BITTEL, K.
1981 Religion und Kult. In Bittel, K., W. Kimmig and S. Schiek (eds.) *Die Kelten in Baden-Württemberg.* Stuttgart: Konrad Theiss Verlag,. pp. 85-117.

BITTEL, K., W. KIMMIG and S. SCHIEK (eds.)
1981 *Die Kelten in Baden-Württemberg.* Stuttgart: Konrad Theiss Verlag.

BROWN, J. A., R. W. WILLIS, M. A. BARTH and G. K. NEUMANN (eds.)
1967 *The Gentleman Farm Site La Salle County, Illinois.* Springfield: Illinois State Museum, Report of Investigations 12.

BURMEISTER, S.
2000 *Geschlecht, Alter und Herrschaft in der Späthallstattzeit Württembergs.* Münster: Waxmann.

DÄMMER, H.-W.
1974 Zu späthallstattzeitlichen Zweischalennadeln und zur Datierung des Frauengrabes auf der Heuneburg. *Fundberichte aus Baden-Württemberg* 1: 284-292.

DIETLER, M.
1995 Early "Celtic" Socio-political Relations: Ideological Representation and Social Competition in Dynamic Comparative Perspective. In Arnold, B. and D. Blair Gibson (eds.) *Celtic Chiefdom, Celtic State,* New Directions in Archaeology. Cambridge: Cambridge University Press. pp. 64-71.

1996 Feasts and Commensal Politics in the Political Economy: Food, Power and Status in Prehistoric Europe. In Wiessner, P. and W. Schiefenhövel (eds.) *Food and the Status Quest: An Interdisciplinary Perspective* Oxford: Berghahn Books.

DIETRICH, H.
1998 *Die hallstattzeitlichen Grabfunde von Heidenheim-Schnaitheim.* Stuttgart: Konrad Theiss Verlag.

EGGERT, M. K. H.
1988 Riesentumuli und Sozialorganisation: Vergleichende Betrachtungen zu den sogenannten "Fürstenhügeln" der späten Hallstattzeit. *Archäologisches Korrespondenzblatt* 18: 263-274.

FISCHER, F.
1982 Frühkeltische Fürstengräber in Mitteleuropa. In *Antike Welt: Zeitschrift für Archäologie und Kulturgeschichte* 13: 1-72. Feldmeilen: Raggi Verlag.

FRANKENSTEIN, S. and M. J. ROWLANDS
1978 The Internal Structure and Regional Context of Early Iron Age Society in Southwest Germany. *Institute of Archaeology Bulletin* 15: 73-112.

GEERTZ, C.
1980 *Negara: The Theater State in Nineteenth-Century Bali.* Princeton: Princeton University Press.

GERSBACH, E.
1969 Heuneburg-Außensiedlung-jüngere Adelsnekropole: Eine historische Studie. In *Fundberichte aus Hessen*, Festschrift für Wolfgang Dehn, Beiheft 1, Marburger Beiträge zur Archäologie der Kelten. pp. 29-34.

1989 *Ausgrabungsmethodik und Stratigraphie der Heuneburg.* Heuneburg Studien VI. Römisch-Germanische Forschungen 45, Mainz.

HERODOTUS
1998 *The Histories.* Translated by Robin Waterfield. Oxford: Oxford University Press.

HINGLEY, R.
1996 Ancestors and Identity in the Later Prehistory of Atlantic Scotland: The Reuse and Reinvention of Neolithic Monuments and Material Culture. *World Archaeology* 28(2): 231-243.

HOLTORF, C. J.
1998 The Life-Histories of Megaliths in Mecklenburg-Vorpommern (Germany). *World Archaeology* 30(1): 23-38.

JOFFROY, R.
1962 *Le Trésor de Vix: Histoire et Portée d'une Grand Découverte.* Paris: Fayard.

KAN, S.
1989 *Symbolic Immortality: The Tlingit Potlatch of the Nineteenth Century.* Washington D.C.: Smithsonian Institute Press.

KELLER, F.
1871 Fünfbühel zu Zollikon unweit Zürich. *Heidengräber 20. Anzeiger für schweizerische Alterthumskunde 1871* 3: 257-261.

KIMMIG, W.
1983 *Die Heuneburg an der oberen Donau.* Führer zu archäologischen Denkmälern in Baden-Württemberg. Stuttgart: Konrad Theiss Verlag.

KRÄMER, W.
1966 Prähistorische Brandopferplätze. In Degen, R., W. Drack and R. Wyss (eds.) *Helvetia Antiqva: Festschrift Emil Vogt.* Zürich: Conzett & Huber. pp. 111-122.

KUHN, D. W.
1987 Fire Ceremony Highlights Indian Reburial. *Ohio Archaeologist* 37(4): 26-27.

KURZ, S.
1997 *Bestattungsbrauch in der westlichen Hallstattkultur.* Münster: Waxmann.

2001 Siedlungsforschungen im Umland der Heuneburg: Fragestellung und erste Ergebnisse. In *"Paläoökosystemforschung und Geschichte": Beiträge zur Siedlungsarchäologie und zum*

Landschaftwandel. DFG-Graduiertenkolleg 462. Universitätsverlag Regensburg GmbH, Regensburg.

2002 Siedlungsforschungen bei der Heuneburg, Gde. Herbertingen-Hundersingen, Kreis Sigmaringen: Zum Stand des DFG-Projektes. *Archäologische Ausgrabungen in Baden-Württemberg 2001*: 61-63

. 2005 Introduction: The Heuneburg in the Course of History. Electronic document, http://www.dhm.de/museen/heuneburg/ed/einf _frame.html, February 28, 2005.

KURZ, S. and S. SCHIEK (eds.)

2002 *Bestattungplätze im Umfeld der Hueneburg.* Stuttgart: Konrad Theiss Verlag.

LUNGU, V.

2002 Tombs, Texts and Images: Fire in Greek Funeral Rituals. In Gheorghiu, D. (ed.) *Fire in Archaeology*. BAR International Series 1089. Oxford: BAR Publishing. pp. 133-142.

MEGAW, J. V. S.

1966 The Vix Burial. *Antiquity* 40: 38-44.

MORRIS, I.

1987 *Burial and Ancient Society*. Cambridge: Cambridge University Press.

MURRAY, M. L.

1992 The Archaeology of Mystification: Ideology, Dominance, and the Urnfields of Southern Germany. In Sean Goldsmith, A., S. Garvie, D. Selin and J. Smith (eds.) *Ancient Images, Ancient Thought: The Archaeology of Ideology*, Calgary: The University of Calgary Archaeological Association. pp. 97-104.

OTTE, M.

2002 Fire as an Evolutionary Factor. In Gheorghiu, D. (ed.) *Fire in Archaeology*, BAR International Series 1089. Oxford: BAR Publishing. pp. 7-9.

RASSHOFER, G.

1998 *Untersuchungen zu metallzeitlichen Grabstelen in Süddeutschland*. Rahden/Westf: Verlag Marie Leidorf GmbH.

REIM, H.

1988 *Das keltische Gräberfeld bei Rottenburg am Neckar: Grabungen 1984-1987*. Archäologische Informationen aus Baden-Württemberg 3, Stuttgart.

2002 Siedlungsarchäologische Forschungen im Umland der frühkeltischen Heuneburg bei Hundersingen, Gemeinde Herbertingen, Kreis Sigmaringen. In *Heimat- und Altertumsverein Heidenheim an der Brenz e. V. Jahrbuch 2001/2002*, pp. 12-33.

RIECKOFF, S. and J. BIEL (eds.)

2001 *Die Kelten in Deutschland*. Stuttgart: Konrad Theiss Verlag.

RIEK, G.

1962 *Der Hohmichele: Ein Fürstengrab der späten Hallstattzeit bei der Heuneburg*. Heuneburgstudien I. Römisch-Germanische Forschungungen 25, Berlin.

SCHIEK, S.

1981 Bestattungsbräuche. In Bittel, K., W. Kimmig and S. Schiek (eds.) *Die Kelten in Baden-Württemberg*, Stuttgart: Konrad Theiss Verlag. pp. 118-137.

SCHNEIDER, S. A.

2003 *Ancestor Veneration and Ceramic Curation: An Analysis From Speckhau Tumulus 17, Southwest Germany*. Unpublished Master's Thesis, Department of Anthropology, University of Wisconsin-Milwaukee.

SNODGRASS, A. M.

1971 *The Dark Age of Greece: An Archaeological Survey of the Eleventh to the Eighth Centuries BC*. Edinburgh: University Press Edinburgh.

SPINDLER, K.

1983 *Die Frühen Kelten*. Stuttgart: Reclam.

VAN GENNEP, A.

1960 *The Rites of Passage*. Chicago: University of Chicago Press.

ZÜRN, H.

1964 Eine hallstattzeitliche Stele von Hirschlanden, Kr. Leonberg: Vorbericht. *Germania* 42: 28-35.

Roasters from the Early Medieval Hillfort at Stradów, Czarnocin Commune, South Poland, in the Light of the Results of Specialist Analyses[1]

Bartłomiej Szymon Szmoniewski, Andrzej Kielski, Maria Lityńska-Zając, and Krystyna Wodnicka

Introduction

The Early Medieval hillfort in Stradów, site 1, Czarnocin commune, Świętokrzyskie district belongs to one of the most extensive defence works that are known from the territory of Poland. It consists of the hillfort's centre, the so-called *Zamczysko*, and three fortified suburbia: *Barzyńskie, Mieścisko, Waliki* (**Fig. 1**). Altogether, the above mentioned parts of the hillfort cover the area of approx. 25 ha.

The hillfort is located on the Wodzisław Hill, which is one of the mezo-regions of the Nida Basin (Kondracki 1981), which – according to the geomorphological division of Poland – belongs to the Miechów Upland (Klimaszewski 1972).

The south-eastern slopes of the Wodzisław Hill are mostly deforested. Like its flat, open top, they are used for agriculture. On the hill sides and slopes there are cultivated terraces. The soils occurring on that territory developed on loess formations and belong to the most fertile soils on the whole territory of the Nida Basin (Oczoś and Strzelec 1986). They can be easily cultivated and have good physico-chemical characteristics.

In the course of the research conducted on site 1 in the years 1956 to 1963, altogether 221 features were found (Maj 1990: 11 ff.). In some of them, dated to the Early Middle Ages, shards of clay vessels were found: roasters (cf. the catalogue of featutres by H. Zoll-Adamikowa and B. Sz. Szmoniewski, in print). Roasters are basin-like, shallow vessels with moderately to very thick walls. They were made of "lightly" fired clay. As an admixture to the potter's clay, usually chaff and other plant remains were added.

The reconstructed roasters are rectangular or, less frequently, round or oval in shape (cf. Malinowski 1959, 1970). They vary in size, with the length and width reaching over 50 cm. The walls range from 2 cm to 4 cm in thickness, and their heights can reach a dozen centimetres or so (Malinowski 1970, Brzostowicz 2002:

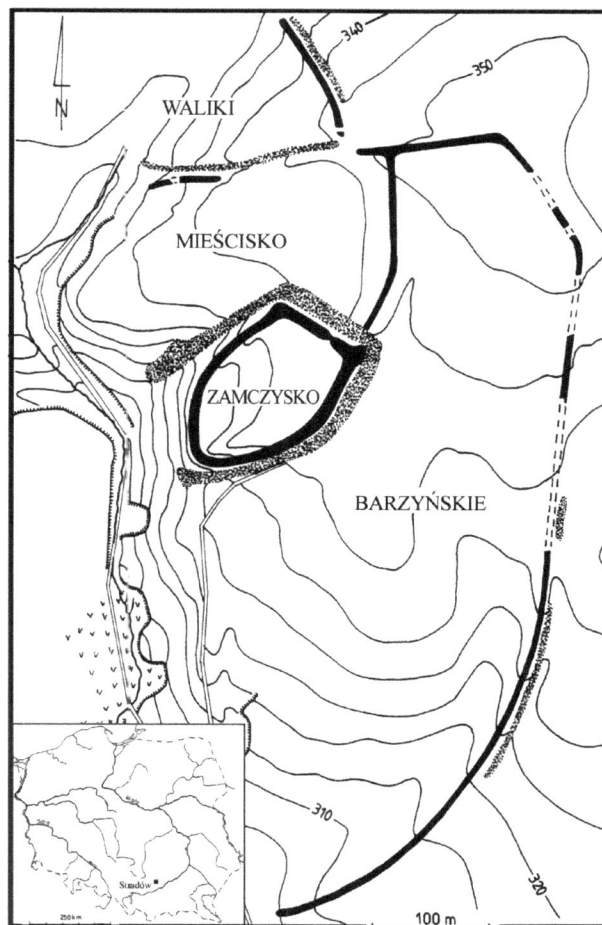

Fig. 1: Stradów, site 1, commune Czarnocin. Plan of the hillfort (hillfort's centre- Zamczysko, suburbia: Barzyńskie, Mieścisko, Waliki).

84). In the archaeological literature on the subject, those vessels were described as typical of the Early Middle Ages (Hersel 1987: footnote 58, cf. dating by Brzostowicz, 2002: 85). On the basis of an analysis of early Slavonic materials from the territory of Poland, M. Parczewski (1988: 75-76) demonstrated that they are not a typical element in the horizon of the early Slavonic culture, they could have been used since the 7th century (Parczewski 1989: 37). It is also difficult to determine the upper time limit of the occurrence of this category of artefacts, although some researchers say that it is the 13th century (Malinowski 1959: 77, 79).

[1] Specialist analyses were made by the Grant of the Scientific Research Commitee (Komitet Badań Naukowych) Grant number KBN 2HO1H02623, *Wczesnośredeniowieczne grodzisko w Stradowie, woj. świętokrzyskie w świetle badań z lat 1956-63 (analizy specjalistyczn*e).

Characteristics of the analysed material

The roasters found the Early Medieval features in Stradów, site 1 are fragmentarily preserved. They are mostly shards of the rims and lumps of the outer parts of the vessels. It must be emphasised that the resemblance that those uncharacteristic pieces have to the lumps of daubed clay does not allow us to determine their exact amount (cf. comments by Malinowski 1959: 68). For the needs of our analysis, only those of the lumps which undoubtedly represent the remains of roasters were used for the macroscopic analysis of plant remains.[2]

A detailed analysis of the preserved forms of the roasters allows us to distinguish two variants among them.

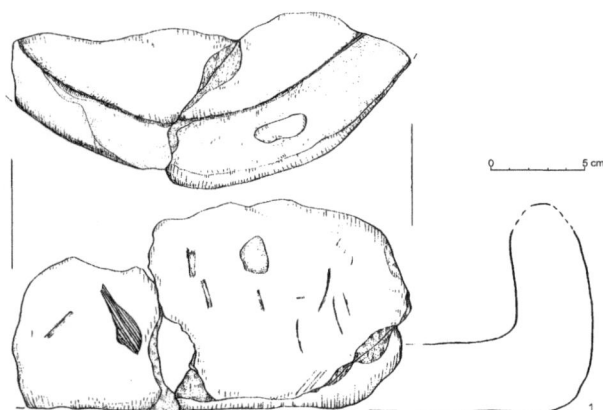

Fig. 2: Stradów, site 1, com. Czarnocin, roaster from feature 25 (variant I). Drawn by E. Osipowa

The first one is characterised by squat forms (cf. **Fig. 2 and Fig. 5A**). The edges of those vessels are gently rounded, while the walls get wider towards the bottom. The outer surfaces are usually brick-red or brown in colour, while the inner surfaces are black. The rims are gently rounded. Their structure is not very solid, and – if touched on the surface – they crumble. Both on their surfaces and in the breaks there are numerous plant impressions (those are mainly impressed blades of grasses). In addition to the admixture of organic remains, there is also some sand and small pieces of marl.

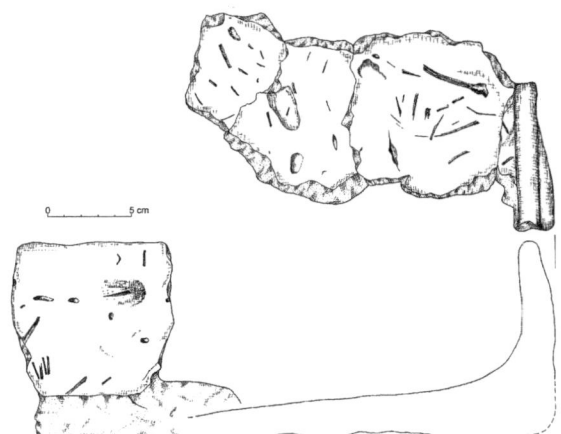

Fig. 3: Stradów, site 1, com. Czarnocin, roaster from feature 25 (variant II). Drawn by E. Osipowa

Variant II is characterised by a more solid structure of the surface (cf. **Fig. 3, Fig. 4 and Fig. 5B**). The vessels themselves were made with greater care. There were plant remains, sand, small pieces of marl and potsherds mixed with the clay. In a majority of cases, their rims are gently profiled and the width of their walls is even. There are only a few fragmentary pieces with the walls lightly protruding outwards.

Fig. 4: Stradów, site 1, com. Czarnocin, roaster from feature 3/M (variant II). Photo by B. Sz. Szmoniewski.

There is a smaller amount of organic remains mixed with the clay than in the case of variant I. The remains are mostly found in the lower parts of the vessels.

Characteristics of the analysed samples

Roasters are among characteristic kinds of the "non-domestic" pottery, found on Early Medieval sites. Owing to the fact of their fragmentary preservation and their similarity to lumps of daubed clay, it is not possible to estimate very accurately their quantity (cf. Malinowski 1959, 69).

---

[2] In the present paper, the result of the analyses of the roasters from the hillfort's centre (*Zamczysko*) will be presented. The remaining materials from the hillfort's suburbia will be presented in the prepared volume of a monography on Stardów, ed. A. Buko. Thus far, only the shards of a roaster from feature 3/M (suburbium *Mieścisko*) have been analysed. It is the best example of variant II among the roasters found on site 1 in Stradów.

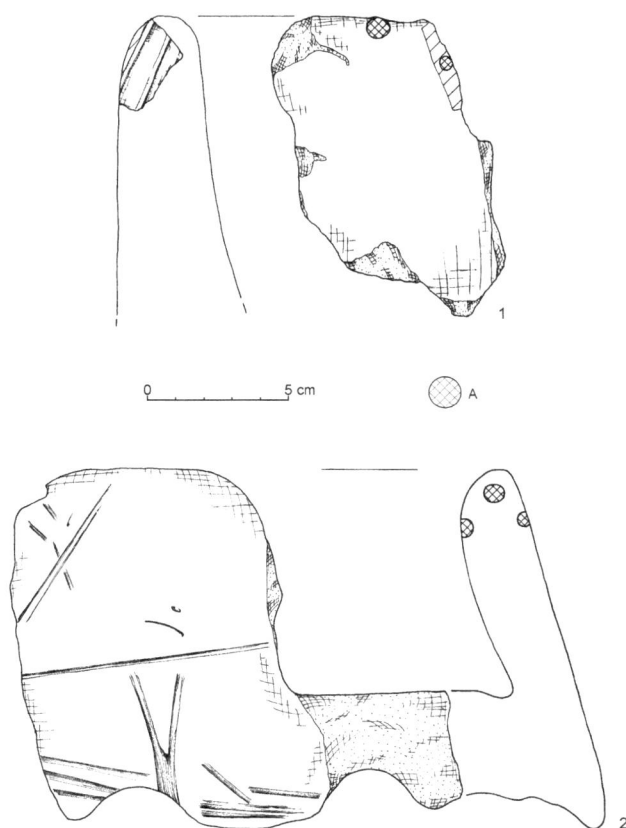

Fig. 5: Stradów, site 1, com. Czarnocin, A) roaster from feature 10; sample no 1 (variant I). Drawn by E. Osipowa, B) roaster from feature 3/M; sample no 4 (variant II). Drawn by E. Osipowa.

Description of samples

We based our investigation of the surface morphology and chemical composition of the roasters on four samples. For the purpose of the investigation, the following measurements were performed: S_{BET} Analysis,[3] absolute density,[4] average density and porosity percent,[5] and X-ray diffraction (XRD).

---

[3] BET surface area (SBET) was measured by means of a multifunction apparatus for measuring BET surface and porosity – model ASAP 2010, manufactured by the American company Micrometritics. BET surface area (SBET) was determined by means of the method of physical adsorption of nitrogen at 77K using the Brunauer-Emmet-Teller equation (theory of a 2-parameter multilayer adsorption). For the calculations, the data of the isotherm of adsorption at relative pressure p/po ranging from ca. 0.06 to ca. 0.20 was used.
[4] Absolute density was measured using the helium pycnometer AccuPyc 1330 manufactured by the American company Micrometrics. With the help of pure helium, the volume of the investigated samples was determined, taking into account standard deviation, and the results were then used to calculate the density. There were 5 measurements conducted in parallel for each sample. Before taking the measurements, the samples were initially desorbed by being rinsed 10 times in pure helium.
[5] The measurements of average density and porosity percentage were made by means of the density analyser GeoPyc model 1360 manufactured by the American company Micrometrics. It is a modern,

For the purpose of analysis, two samples per each of the two variants of roasters were chosen. They were the most typical examples of the variants.

Variant I
Sample No. 1.

The roaster's sherd came from feature no. 10 (**Fig. 5A**).

The sample was taken both from the external internal surfaces of the vessel. They were fired with a varying degree of intensity, and they also differed in their surface colour. In the roaster's break, a sherd of domestic pottery was found. The roaster itself was only lightly fired. Its break had two colours: the external surface was red, while the core was dark, almost black, because of the occurrence of unoxidized carbon. The potter's clay was probably made from very rich clay, and therefore had to be "thinned" by means of adding a large inclusion of organic matter such as chaff or husks. There were also inclusions of small quantities of sand, grains of limestone or broken pottery pieces. The surface of the vessel crumbled if pressed with a finger.

Sample No. 2.

The roaster came from feature no. 25 (**Fig. 2**).
The sample was taken from the core of the vessel. The roaster's surface was not very cohesive and it had a tendency to crumble. The vessel was only lightly fired. The break showed no diversity of colour – it was uniformly almost black. There were only a few specks of brick-red. The potter's clay contained a very large inclusion of organic matter in the form of chaff, husks and a small quantity of limestone grains. On the roaster's surface, impressions of plants could be seen.

Variant II
Sample No. 3.

The roaster came from feature no. 25 (**Fig. 3**).
The analyzed sherd had a very solid structure, and was clearly different from other roasters. The external surface of the vessel was brick-red in colour. The colour of the core was dark brown, while the interior surface (towards the inside of the vessel) was light red with traces indicating that it must have been covered with a thin layer of light-brown clay. The potter's clay was mixed with sand and a large quantity of plant remnants (in this case it was chaff). On the above-mentioned layer, which must have been created as a result of covering the sides of the vessel with clay, there were very numerous impressions of grass. Here and there, small fragments of the clay layer had fallen off. For the microscopic analysis, three

---

fully automatic device with a microprocessor. The measurement consists in determining the "outer" volume (including the pores: the so-called "envelope volume") of a sample. The device performs this by determining the volume of an appropriately chosen amount of powder (blank run), and then – of the volume of the powder together with the sample (sample run).

samples were collected, two from the external surface, and one from the core.

Sample No 4.

The roaster came from feature no. 3 in Mieścisko suburbium (**Fig. 4; Fig. 5B**).
This roaster's shard had a very cohesive, solid structure, not susceptible to crumbling. The break was of two colours. The external layer was lighter in colour – light brick-red, and the core was darker. On the surface, there were a few specks of rusty colour, probably traces of "fire". There was an organic inclusion, and a small quantity of sand. On the roaster's surface, there were impressions of plants.

Summary of the result of analysis of the roaster samples

Roaster I (the pottery inclusion). This sample is distinguished by high porosity, and – as a consequence – by low average density. The dominant phase constituent is quartz, accompanied by albite, microcline and illite (see Tables 1 and 2).

Roaster I (the sample from the interior surface). This sample is different from the sample obtained from the wall of the vessel, because its porosity is relatively low and the result of SBET Analysis is higher. The phase constituents are also different. In this sample, quartz is accompanied by calcite and albite. The two samples (from the wall and from the bottom of the vessel) differ significantly; and it is also confirmed by their varying densities (see Tables 1 and 2).

Roaster II (the core sample). This sample is characterised by high porosity and a high result of SBET Analysis. In addition to quartz, there also are: illite, albite and microcline. That indicates that the roaster was fired relatively "lightly" (see Tables 1 and 2).

Roaster III (the external surface, core and interior surface). Also these samples have relatively high porosity and low SBET Analysis results. Characteristically, the results are not very different between the three samples. Also their phase constituents are similar, with quartz dominating and accompanied by illite, microcline, anorthite and albite. The SBET Analysis and density measurements are also similar. It shows that the samples are homogenous and that the vessels were fired at a relatively low temperature (see Table 1 and 2).

Roaster IV (the external surface). This sample is characterised by a slightly lower porosity and a significantly lower SBET Analysis results. It could be due to the occurrence of albite, which accompanies quartz as the second most prevailing constituent (see Tables 1 and 2).

Table 1: Results of measurement the Bet-Surface Area, Density and Porosity of Roasters

Sample Type	S_{BET} m^2/g	Absolute Density g/cm^3	Average Density g/cm^3	Specific Pore Volume cm^3/g	Porosity Percent %	Mezo micro pore cm^3/g
Roaster I	3.00	2.7337	1.4277	0.3345	47.8	0.0075
Roaster I (pottery inclusion)	7.50	2.5516	1.8222	0.1568	28.6	0.0259
Roaster II	4.87	2.6278	1.3978	0.3343	46.8	0.0140
Roaster III (internal surface)	–	2.6889	1.6210	0.2449	39.7	–
Roaster III (core)	5.41	2.7066	1.7492	0.2022	35.4	0.0136
Roaster III (external surface)	4.71	2.6992	1.6520	0.2348	38.8	0.0167
Roaster IV (internal surface)	1.43	2.5859	1.7155	0.1962	33.7	0,0036

Table 2: Results of measurement X-Ray diffraction XRD of the roaster

Sample Type	Common : 1 (Scan 1)	Scan Axix: 2:1 sym
Roaster I	Quartz	Albite, microcline, illite
Roaster I (pottery inclusion)	Quartz	Calcite, albite
Roaster II	Quartz	Illite, albite, microcline
Roaster III (internal surface)	Quartz	Anorthite, illite, microcline
Roaster III (core)	Quartz	Illite, microcline, albite
Roaster III (external surface)	Quartz	Illite anorthite, microcline
Roaster IV (internal surface)	Quartz	Albite, illite, microcline
Roaster IV (external surface)	Quartz	Albite, microcline, illite

Summary of the botanical analyses

Botanical analysis was conducted on all of the roasters' fragments that contained remnants of plants. The pieces chosen for analysis mostly came from the "hillfort's centre". Altogether, we analysed 121 pieces from the "hillfort's centre", which came from 24 samples from 13 features (see Table 3).

The analysed pieces of clay contained remnants of plants both on their external and internal surfaces. Most abundant remnants belongs to cereals and/or to different species of wild grasses *Cerealia indet. vel Gramienae indet.* (Table 3). There were impression largest or smaller charred fragments of grain leaves and stems. In burnt pieces were found two charred caryopses of bread/club wheat *Triticum aestivum* s.l. There were also preserved two impressions and more numerous charred remnants of grains of an undetermined wheat *Triticum* sp. In the

Table 3: The results on investigation of plant imprints on roasters from the Early Medieval fort at Stradów, site 1.

cereal	type	feature	number of features												
taxa name	kind of remains	Preservation	7	10	18	19	21	25	44a	54	57	61	73	78	88
Triticum aestivum s.l.	z	sp										2			
Triticum sp.	z	od										5			
		sp													16
Hordeum vulgare	z	od	1												
Panicum miliaceum	z	sp										2			
Secale cereale	z	od						2							
Cerealia indet. vel	sł	od	++	++	15	+		6	2	4	+	5	+	+	+
Gramineae indet.		sp	++	++	6	3	+	+					+		+
Fallopia convol-vulus	o	sp						1							
Bromus sp.	z	sp													4
Gramineae indet.	z	sp										2			
Double leaf	leaf	od						1							
Un-determined	o	od						1							

Explanations: kind of remains: sł – chaff, z – caryopsis, o – fruit; preservation: od – impressions, sp – charred remain inside imprints; number of specimens: + – numerous remains.

sample collected from feature no. 88, there were numerous remnants of wheat. In addition, there was one impression of common barley *Hordeum vulgare* L. and charred caryopses of common millet *Panicum miliaceum* L., which were present in feature 61.

Among wild herbaceous plants, there were preserved charred caryopses of brome grass *Bromus* sp. and undetermined species of wild grass *Gramineae* indet. In the sample from feature no. 25, were found some fragmentary pieces of the fruit of black-bindweed *Fallopia convulvulus* (L.) Á. Löve and one, badly preserved, trace of a leaf of a dicotyledonous plant. In sample (no. 147/57) from feature no. 25 were observed a single impression of an undetermined fruit with clearly marked longitudinal "ribs" on the surface (**Fig. 6**).

Discussion

Our research has confirmed that roasters were fired at relatively low temperatures, not exceeding 750° C. We have also found large quantities of organic matter (in the form of plant remnants), which were used for "thinning" the rich clay used for making roasters.

In all of the samples, the X-ray diffraction shows quartz as the prevailing constituent, accompanied by minerals such as illite, albite, anorthite and microcline. In spite of the similarity of their phase constituents, the quality of the samples differs significantly. Thus we can distinguish 2 variants of roasters. Type I is characterized

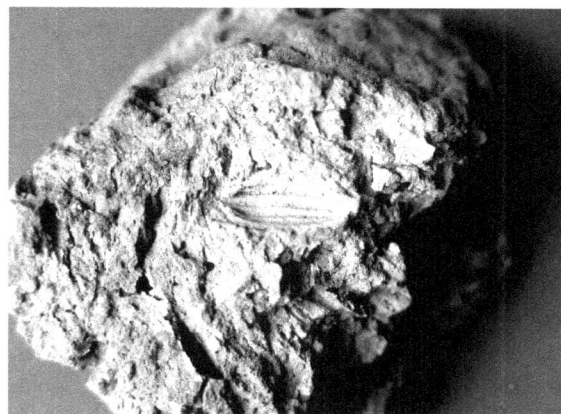

Fig. 6: Impression of an undetermined fruit from the roaster from feature 25. Photo by B. Sz. Szmoniewski

by considerable brittleness, little mechanical strength (the pieces crumble easily if the break is pressed with a finger), and large inclusions of organic matter. What we can observe is confirmed by analysis of porosity and average density: the roasters' porosity is high, and their average density is low. On the other hand, the roasters of variants II are much less brittle and more solid. The inclusions of organic matter that they contain are smaller. As a result, their porosity is lower and their average density higher.

Variant I was represented by samples 1 and 2, while variant II by samples 3 and 4, particularly by their interior fragments. Usually, roasters of type I are not so well fired, and their surfaces are not solid and they crumble easily. Roasters of type II are better fired and their surfaces are more solid. In one instance, the surface was covered with a thin layer of clay.

Analysis of broken pottery inclusions in the clay used for making a roaster in sample 1 clearly shows the difference between domestic pottery and other kinds of pottery in respect of both porosity percent and X-ray diffraction.

Attention should also be drawn to the fact that the cores of the samples that can be seen on the surfaces of the breaks are, in many cases, noticeably darker in comparison to the external surfaces of the vessels. This is due to the occurrence of carbon from decomposing organic matter, which did not oxidize during the process of firing the roasters.

Analysis of the morphology of the samples' surfaces by means of the scanning electron microscope equipped with the attachment for chemical microanalysis showed the occurrence of titanium and zirconium in samples III and IV. Titanium dioxide usually accompanies aluminium oxide in alumino-silicates, while the occurrence of zirconium is difficult to explain at this stage of our research. The occurrence of titanium and zirconium has also been observed among vessel pottery. This fact may indicate that the clay used for vessel pottery and roasters came from the same sources.

Roasters, and particularly their function and the possible ways in which they could have been used, have long been the subject of discussion among researchers dealing with the problems of the Early Middle Ages (cf. Malinowski 1959; 1970: see further literature ibid; Parczewski 1988: 75). It has been proposed that roasters served as containers for rinsing, for salt evaporating (Malinowski 1959: see literature overview ibid), or served some auxiliary function in metallurgy (Szafrański 1961: 51; Zoll-Adamikowa 1979: footnote 101). The majority of researchers seem to support the view that those vessels were used in the processes connected with chaff removing and roasting of corn (cf. Malinowski 1970: 332; Parczewski 1988: 75). Nowadays, mainly owing to the descriptions and reconstruction by K. Moszyński (1967: 265, fig. 223, cf. reconstruction in fig. 6), it is assumed that roasters were mainly used in the processes of removing chaff from corn or in bread baking (Brzostowicz 2002: 85, see: previous literature on the subject ibid).

The thesis of roasters being used as vessels during thermal processing of grain or grain products seems to be even more justified by the needs of natural economy. The grain, which has been subjected to high temperatures, can be better stored than the grain which has only been left to dry. Prepared in this way, grain could be stored in pits for a considerable length of time, and – what is extremely important – it did not lose its good taste (Podkowińska 1964: 31 ff., Maurizio 1926: 202, 203). However, the grain that is subjected to thermal processing loses its power to germinate; therefore it does not seem possible that the whole supply was exposed to high temperatures before storage. Thermal processing could have only been conducted just before the grain was used for consumption.

It is worth considering here the way in which roasters were fitted into the "heating unit" (or "heating device"). In the hillfort of Stradów, roasters were found in the features which functioned as areas where housework or farming work was done, or which were rubbish dumps as well as those that were strictly speaking human dwellings (**Fig. 7**, cf. Zoll-Adamikowa and Szmoniewski, in print). In several cases, they were found in the close vicinity of kilns. In the archaeological literature on the subject, it is accepted that roasters were not made to be carried (Parczewski 1988: 75; Parczewski 1989: 37). The traces of thin poles or maybe branches found on some of the shards of the roasters' bottoms have been interpreted as the remains of wooden supports on which they were placed while being moulded (Malinowski 1959: 73, 74). Such indentations were also noticed near the side edges, and they were found along the longer side of a vessel, on a few shards from Stradów. The question arises whether those thin poles could have been used to facilitate carrying of the vessels. Perhaps, the find of a "corner"

shard of a roaster from Obiszów, site 9, Grębocice commune, with a preserved wooden element may serve as some evidence in favour of this hypothesis. It could have been used for carrying the vessel. The shard was found in the fill of feature 26 B (Baron and Rzeźnik 1999: 270, fig. 1). That assemblage should be dated to the 10th century, and more precisely to 2nd - 3rd quarter of the 10th century. It seems probable that the vessels were put onto the "heating units" (e.g. kilns) when grain needed to undergo thermal processing.

The roasters found in Romanian settlements, which were dated to the 6th, 7th century, were fitted onto clay kilns (cf. Dolinescu - Ferche 1995). They were to serve as lids, placed on the upper part of a kiln (cf. reconstructions: cf. Dolinescu - Ferche 1995, fig. 1; fig. 2; fig. 5; fig. 6; fig. 10; fig. 11; fig. 13; fig. 14; fig. 17; cf. reconstructions in fig. 7).

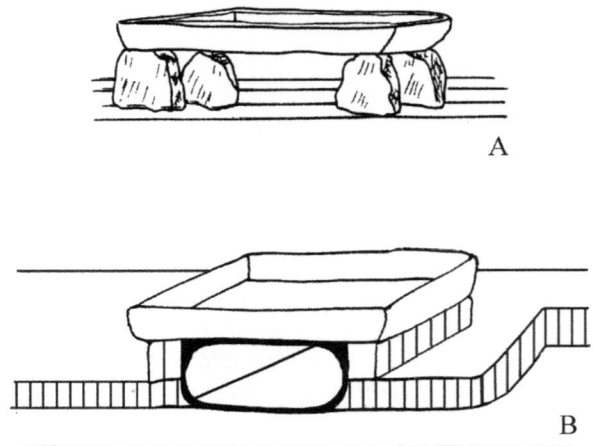

Fig. 7: Reconstruction of the way of mounting roasters onto the "heating devices". a) after Dolinescu-Ferche 1995; b) after Moszyński 1967

The roasters from the sites on the territory of Poland, similarly like on other Slavonic territories, are a very frequent find among vessel pottery. Their form and the admixtures added to the clay are almost identical with the shards known from other sites (cf. Malinowski 1959). The diversification noticeable among the roasters from Stradów may reflect various functions of those vessels. At the present stage of research, it is however difficult to determine exactly the functions of the two variants known from Stradów. Apparently, they were used for roasting or drying grain.

Acknowledgements

The authors would like to thank Dr Paweł Rzeźnik for his permission to include the information about the roaster from Obiszów.

Bibliography

BARON, J. and P. RZEŹNIK
1999 Wczesnośredniowieczny budynek z tzw. korytarzykiem wejściowym z osady w Obiszowie na Wzgórzach Dalkowskich, *Śląskie Sprawozdania Archeologiczne* 41: 269-280.

BRZOSTOWICZ, M.
2002 Bruszczewski zespół osadniczy we wczesnym średniowieczu, Poznań.

DOLINESCU-FERCHE, M.
1995 Cuptoare din interiorul locuinţelor din secolul al VI-lea e.n. de la Dulceanca, *Studii şi Cercetări de Istorie Veche şi Arheologie*: 46(2): 161-191

HENSEL, W.
1987 Słowiańszczyzna wczesnośredniowieczna. Zarys kultury materialnej, wyd. IV, Warszawa.

KLIMASZEWSKI, M.
1972 Podział morfologiczny Południowej Polski, *Czasopismo Geograficzne* 17 (3).

KONDRACKI, J.
1981 *Geografia fizyczna Polski*, Warszawa

MAJ, U.
1990 Stradów, stanowisko 1. Część 1. Ceramika wczesnośredniowieczna, Kraków.

MALINOWSKI, T.
1959 Wczesnośredniowieczne prażnice w Wielkopolsce, *Przegląd Archeologiczny* 9: 68-80.
1970 Prażnice. *Słownik Starożytności Słowiańskich* IV(1): 332, Wrocław-Warszawa-Kraków.

MAURIZIO, A.
1926 *Pożywienie roślinne i rolnictwo w rozwoju dziejowym*, Warszawa

MOSZYŃSKI, K.
1967 *Kultura ludowa Słowian*, v. 1, kultura materialna, II ed., Warszawa

OCZOŚ Z. and J. STRZELEC
1986 Gleby Niecki Nidziańskiej. *Studia ośrodka Dokumentacji Fizjograficznej*, 14: 311-331.

PARCZEWSKI, M.
1988 Początki kultury wczesnosłowiańskiej w Polsce. Krytyka i datowanie źródeł archeologicznych, *Prace Komisji Archeologicznej PAN*, Kraków.
1989 *Żukowice pod Głogowem w zaraniu średniowiecza*, Głogów: Głogowskie Zeszyty Naukowe.

PODKOWIŃSKA, Z.
1964 Spichrze ziemne w osadzie kultury pucharów lejkowatych na Gawrońcu-Pałydze w Ćmielowie, pow. Opatów, *Archeologia Polski* 6 (1): 27-45.

SZAFRAŃSKI, W.
1961 Wyniki badań archeologicznych w Biskupinie pow. Żnin, na stanowisku 6. In Szafrańscy, W. i Z. (ed.) *Z badań nad wczesnośredniowiecznym osadnictwem wiejskim w Biskupinie*, Wrocław, pp. 7-144.

ZOLL-ADAMIKOWA, H.
1979 *Wczesnośredniowieczne cmentarzyska ciałopalne Słowian na terenie Polski. v. II. Analiza.* Wrocław: Wnioski.

ZOLL-ADAMIKOWA, H. and B. SZ. SZMONIEWSKI
in print Tabelaryczny opis obiektów ze Stradowa, stan. 1, gm. Czarnocin, woj. Świętokrzyskie. In Buko, A. (ed.), H. Zoll-Adamikowa, B. Sz. Szmoniewski, A. Tyniec-Kępińska, and M. Wołoszyn, *Wczesnośredniowieczne stanowisko w Stradowie.* Kraków: A. Buko.

Pyrotechnology and Local Resources in Chianti Shire: From Clay, Limestone and Wood to Bricks, Lime and Pottery Making. Some Preliminary Notes

Marta Caroscio

Abstract

The aim of this paper is not only to draw an account on the research status concerning pyrotechnology in Chianti shire, but also to try to understand how social and economic conditions can influence the exploitation of natural resource. Focus will be on different technical devices connected to the use of fire, the archaeological evidence will be a starting point, but written[1] and oral sources, as well as the evidence deriving from place names, will be taken into account. A systematic study has been undertaken only recently; in fact, the research is at a starting point; the archaeological record which will be presented derives mainly from surveys, samples and emergency excavations done by local group of volunteers supervised by the Soprintendenza (Heritage). Thanks to the data they collected it was possible to start a broader research project aimed at locating the Medieval and Modern manufacturing sites. The remains of production structures has been documented and studied, taking into account raw material supplies too. As the road system in Tuscany made carriage costs very high, markets were usually organised on a local base and goods tent to be transported only within a few miles (Quirós Castillo 2003: 397).[2] In this context clay, limestone and wood needed to be in the nearby of the manufacturing place, so that transports were reduced as much as possible (Goldthwaite1980: 178). According to that, both the location of the resources and the environment should be regarded as key elements to understand why manufacturing activities were concentrated in certain areas. The research is still at a starting point; nevertheless, it is possible to draw an account on the methods used, which involves the use of different sources including ethno-archaeology, giving some preliminary results. Firstly, the historic background linked to manufacturing activities will be discussed, showing how the characteristic of landscape and the location of natural resources allowed the development of pyrotechnology in different crafts activities. Archaeological evidence of workshop structures will be presented considering that in some cases bricks and lime making as well as pottery manufacturing could coexist in the same site (Goldthwaite1980: 186-7). In doing so written sources will be taken into account. The discussion will consider the possibilities of developing the research as well as the question which are still to be answered regarding the distribution of manufacturing sites and the transmission of technical skills.

1. Production and manufacturing models

Before analysing crafts linked to pyrotechnology, it is worth considering the context in which these activities took place, trying to reconstruct some producing models. In doing so, the links between the locations of kiln sites and the supplies of raw materials will be taken into account.[3]

When describing kilns and working premises, focus will be on understand if different approaches in the use of fire resulted into different achievements. Some hypotheses on the producing model present in the sub-regional area considered (Florentine Chianti shire) have been done on the base of the data available and making comparisons

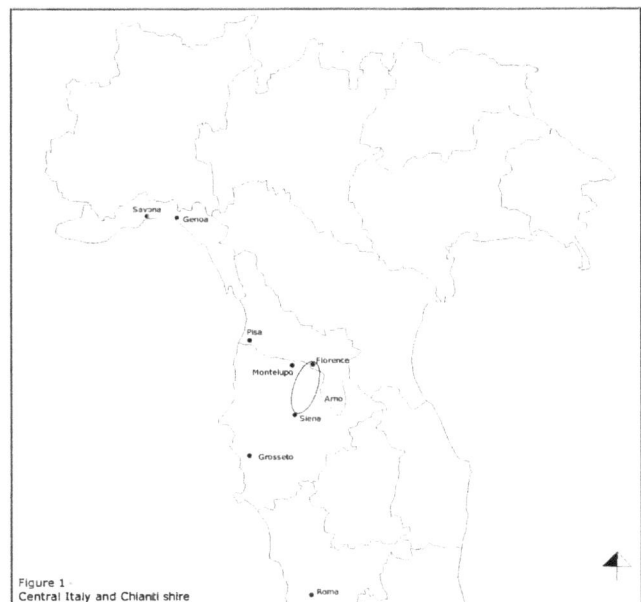

Figure 1
Central Italy and Chianti shire

Fig. 1: Central Italy highlighting Chianti shire.

with other areas, but further researches, both on archive sources and on archaeological evidence are necessary. Studies on the Seines Chianti shire and the environs of

---

[1] Unfortunately, manorial account are lacking, but there are statement of tax paying, municipal regulations and technical treaties, each providing different kind of information on the manufacturing process and on the quality of products. As it will be discussed below, from the 14th century onwards, Communes started regulating dimensions and prices of artefacts in order to avoid frauds and the presence of monopolies. In other places, like in Britain, manorial account are widely present (Moorhouse 1988: 33)

[2] As it will be discussed below the production was mainly on a local scale (Benente and Biagini 1990; Restagno 1980; Ciciliot 1985; Albisola 1975).

[3] The relationship existing between wood, clay and water is fundamental in the manufacturing process and could influence the costs of final products as well (Covino and Giansanti 2002: 13).

Siena, on northern Tuscany and Liguria, on Florence (**Fig. 1**) and its economic producing system as well as research on other European countries (Darvill and McWhirr 1984; Rice and Kingery 1997) have shown some similarities which made possible reconstructing a "model" likely to be applied to this context.

1.1 Historic and environmental background: economy and materials' supplies

The study of bricks, lime and pottery manufacture is important not only in itself, as giving essential information on the development of building techniques, but can be used as well as a way of reconstructing some aspects of the late medieval society which goes behind the analyse of mansions making and pyrotechnology. In the past only a few studies have taken into account the importance of the economic background in the understanding of the late medieval manufacturing system (Goldthwaite 1980: 171-209), but things have been changing in the last two decades and recent works have focused on this point (Quirós Castillo 1996, 2003; Rauty 1987). Some attempts in reconstructing a producing model has been done not only for sub-regional areas such as Siena (Balestracci 2000) and northern Tuscany (Quirós Castillo 1996) but for the whole Tuscany (Quirós Castillo 2003) and for other regions like Liguria as well (Ferrando Cabona and Mannoni 1988). Differences in the use of various materials are distinctive signs of the characteristic of the producing system operating in a certain period (Wickham 1988: 119), being the availability of natural resources the central point (Quirós Castillo 1996: 47; Carnasciali and Roncaglia 1986: 22).

The beginning of the 11th century was the starting moment for bricks manufacturing in Tuscany; as far as now the research has shown that the first kilns where the craft started, were those located in the Florentine Chianti shire. The origin of this new activity can be understood only if considered in relationship with the economic and the social background it originated in. In Florence[4] the first workshop appear only one century later; it has been suggested that a cause for this chronological shift in starting producing building materials between the town centre and the countryside could be the possibility of reusing architectonical elements from older buildings (Parenti and Quirós Castillo 2000: 223-4).

Building renewal became systematic and on a wide scale only from the mid 12th beginning of the 13th century, being more relevant in the last one (Carnasciali and Roncaglia 1986: 7) and probably linked to the expansion of Florence (Vallacchi 1991: 43-4).[5] The general

demographic decrease of the 14th century and the progressively abandon of the countryside in favour of towns had several consequences, among which the crisis in some manufacturing activities. Because of the lack of skilled workers the salary of some artisans, (i.e. stone-masons) increased considerably (Balestracci and Piccini 1977: 149). The unavailability of chiselled stone on the marked resulted not only in the rise of prices, but in the reuse of materials from abandoned or ruined buildings and in employing different ones like bricks.[6] This happened not only in Tuscany, but in other regions as well: in the nearby of the Appennini mountains in Abruzzo, in an area where limestone is widely available, bricks started to be used and were often combined with pebbles and other stones during the 11-14th centuries (Serafini 2003: 165).

As it will be discussed when presenting possible production models, there are differences between towns and countryside, where manufacturing activities are deeply linked to the local economic system. In the countryside bricks and tiles are usually made for monasteries or religious Orders' buildings or are produced for small settlements; as a consequence there might be no uniformity in the making (Parenti and Quirós Castillo 2000: 227). It has been suggested that, when Gentry and Religious Orders' control on agricultural economy was rather limited, the "autonomy" of peasants made possible the creation of small estates; if the appropriate natural resources where present it was then possible to start a manufacturing activity.[7] The most recent studies on Tuscany has shown that despite bricks and tiles manufacturing started in the countryside, it become a stable activity when controlled by towns; there are frequent cases of special agreements between town councils and artisans in order to have a certain amount of products guaranteed (Caldelli 1991: 36).[8]

During the late Middle Ages Florentine kilns were permanent installations, as shown by assessment of tax paying (Goldthwaite 1980: 179); moreover, if we consider the production of bricks referring to the whole economic system in Medieval Tuscany, it is quite evident

---

[4] It should be remembered that kilns were usually kept close to springs or to rivers, being water necessary not only in the manufacturing activity but in reducing the risk of fires. To avoid fires, during the 15th century kilns were progressively moved out of the town walls; in Florence two centuries later they had all been moved in the environs of the town (Goldthwaite 1980: 180).

[5] An increase in number of building and in manufacturing sructures is known in this period all around Europe (Cortonesi 1987).

[6] Some studies on comparing bricks and chiselled stone in the same area during the same period have already been done (Parenti and Quirós Castillo 2000: 235), but it would be worth full having data on the entire region. It is worth mentioning that despite bricks were less expensive than chiselled stone, they still required a great amount of working hours and for this reason raw stone was still employed for building of less importance (Caldelli 1991: 33).

[7] This statement can be applied to any regional area. The presence of clay supplies in the nearby of woods has been studied for Roma and its surroundings as well (Cortonesi 2002: 136).

[8] The Commune bought the largest amount of bricks, but its role in the production is not clear. It might be possible that in some cases it had a monopoly in this respect. Since the beginning if the 14th century the Commune used to fix prices for tiles and bricks as well as for lime (Balestracci and Piccini 1977: 71-2) in order to keep speculation under control (Caldelli 1991: 36). The practice of controlling prices was known in other region of central Italy such as Umbria since the 13th century (Busti and Cocchi 1996: 20). In Rome the Popes wanted to avoid the presence of a monopoly in producing bricks and roof tiles as well as in buying and stocking them (Cortonesi 2002: 124-5).

that the craft of architectonical elements was one of the most relevant activities done in town during the 13th century (Quirós Castillo 2003: 388). This might be linked to the fact that cities in Tuscany stand nearly always on alluvium valleys, an aspect which makes easier the access to clay's supplies.[9] It has been convincingly discussed that there are differences between towns since the 12th century, when important building started to be elevated using bricks: two groups of towns can be identified. In Pisa, Siena and Lucca bricks totally replaced stones, even though supplies could be easily accessed, while in Florence, Pistoia, Prato and Arezzo bricks were used on a more limited scale, even if the production was stable and artefacts were probably traded on a broader market (Quirós Castillo 2003: 393). Nevertheless, it should be remembered that in Renaissance Florence stone took over bricks at a certain stage (Goldthwaite 1980: 171).

The difference is quite evident if we consider public and religious buildings, in fact, while for the first group of towns clay artefacts are used for the whole construction, in the latter they are employed only for some parts (Parenti and Quirós Castillo 2000: 229). Differences in the production scale is not the only one present in Tuscan towns, where there is a great level of variability in bricks dimensions as well; this aspect have been studied and the results achieved show that despite the expansion of Florence and its political control on a regional scale, there was not an unification in measures,[10] being Pisa the only exception (Corsi 1991; Quirós Castillo 1996, Parenti and Quirós Castillo 2000: 226-7).

Being this the case and considering that bricks dimensions were regulated by Communes, by studying them it is possible to elaborate a mathematic model in order to reconstruct different production phases (Mannoni and Milanese 1988).[11] Dimensional variability should be linked not only to production devices in the local area, but related to markets and to the politic control of the territory as well (Quirós Castillo 2003: 395-6, 400). There is a trend in reducing dimensions,[12] which might be related to speculation[13] as bricks were sold by thousands

and craftsmen paid a fix amount of tax for each firing process; by making small bricks they could have higher income and maintain the production costs at the same level (Vaquero Piñeiro 1996: 484; Vaquero Piñeiro 2002: 152). Speculation can not be regarded as the only reason: technical devices should never been neglected; in fact, it is easier to control the firing process of small items rather than that of big objects[14] (Parenti and Quirós Castillo 2000: 222, 230-1).[15]

One of the most important case study of total replacement of expensive materials such as stone in favour of more economic resources like clay is Siena, where the exclusive use of bricks from the mid 14th century[16] onwards (Balestracci 2000: 418) might be linked to the economic background and the salary system of that time (Carnasciali and Roncaglia 1986: 9). Since then, bricks had been the most used among building material in Siena and in the territory under its control, the clay pit being always located in the nearby of the manufacturing structures (Balestracci 2000: 419).[17] Probably because of the wide-scale production and the large variety of different products (bricks, roof tiles, plaster, ceiling tiles etc.)[18] the costs in Siena were the lowest ones in the region (Parenti and Quirós Castillo 2000: 232). The study of place names has shown that there are about 140 settlements named after "kiln" in the environs of Siena (Passeri 1983: 170-2), most of them being concentrated in the zones where clay was widely present (Vallacchi 1991: 42).[19] Some of them are known from written sources and

---

[9] It has been suggested that the origin of the production and the way it spread might show that technical skills were taught by travelling artisans (Parenti and Quirós Castillo 2000: 234).

[10] This can be referred not only to the dimension of bricks but to the whole system of measurement, which was unified by the Great Duke Peter Leopold in 1782 (11th July). Differences in length and wideness of bricks are known in Roman times as well (Biffi 2002 (ed.), 10.8-19).

[11] When studying this aspect it is vital to distinguish intentional from non-intentional variation (Corsi 1991: 28)

[12] The only town where there is an increase in the dimension of bricks is Siena (Corsi 1991: 29). The continuous decrease in the dimensions of bricks from the 12th to the 19th century is generalised not only in Italy, but in other parts of Europe such as Spain and Serbia as well (Quirós Castillo 2003: 394).

[13] To avoid speculations and frauds, public authorities started regulating measures and prices (Zdekauer 1897: 178-180). In Florence the dimensions of bricks were fixed in 1325 as 29x14.5x7.25 cm (Goldthwaite 1980: 208-9); the same happened in Rome (Vaquero Piñeiro 2002) and Siena as well, but in Siena they tended to enlarge instead of getting smaller from the 13th century onwards (Balestracci 2000: 423). The first town in Tuscany to regulate bricks dimension was Pistoia in 1313 and then again in 1328 (Bottari Scarfantoni 2004: 226).

[14] Again we can resemble some similarities in pottery production. The cost of big vessels such as oil jars was quite elevated not only because of the quantity of raw material necessary in making them and for the time employed in the manufacturing activity, but mainly for the high risk of cracking during the firing process. The clay should be very nicely dried and it was quite difficult to keep the temperature constant for a long time in order to allow the fabric cooking nicely in all its parts. For bricks as well the uniformity in cooking was quite important for bricks as well in order to make them resistant to water (Corsi 1991: 24).

[15] It is worth noting that the number of bricks produced in one firing could vary greatly. It has been calculated that during the 17th century in Holland and northern Europe brick kilns could produce up to 600.000/650.000 in one time (Goldthwaite 1980: 176).

[16] Since the beginning of the century (1302) in Siena the Commune ordered to destroy wood and mud houses and to rebuild them in bricks (Lisini 1903: II 406-7). The spreading of the use of bricks in Siena has been studied by Balestracci and Piccini (1977: 64) and then again by Balestracci (2000); the scholars present evidence as well of the change occurred after the beginning of the 14th century, when Public authorities started to forbid building wooden structure and made it compulsory to replace the existing ones. In the same years (1035-6) both in Siena and Lucca the town walls were destroyed to replace the wooden parts with bricks in order to reduce the risks of fire (Bottari Scarfantoni 2004: 225).

[17] The scholar has summarised how the manufacturing activity developed, underlining how, during the second half of the 13th century, kilns were moved outside the town walls. At the same time there were bans forbidding to dig clay in public streets (Lisini 1903: 136).

[18] The kilns belonging to the Hospital of Santa Maria della Scala in Siena was located in the environs of the town, in a small village called Cuna and produced several kinds of bricks and in large amounts (Vallacchi 1991: 45-6). Concerning roof-tiles, archive sources have shown that during the 14th century they were regarded as a luxury product (Vaquero Piñeiro 2002: 142-3).

[19] The area in the environs of Siena known as "Pian delle Fornaci" (kiln area) is mentioned by Repetti (1833-45, II: 239) as well. The growing

had a long-term producing activity, which lasted up to the 18th century (Vallacchi 1991: 45-6; Corsi 1991: 25) but the majority seams to be linked to temporary needs, their production dieing out quite quickly (Vallacchi 1991: 43). Moreover, documents do not states if they were specialised kilns or not,[20] but give evidence for those belonging to the council or not (Balestracci 2000: 421).[21] In the sub-regional area that we have taken into account there are about 10 names referring to quicklime production and even more referring either to bricks, pottery and tiles manufacturing or more generally to the presence of a structure where fire was used.[22]

If this was the case for Siena, it can be said that the 14th century was everywhere a moment for the beginning of a distinctive bricks architecture. While in the early Middle Ages bricks industry had the same importance as in Roman times, in the late Middle Ages it become increasingly important and linked to towns' manufacturing activities as well (Goldthwaite1980: 173-5). Major changes in house construction occurred in Tuscany only in the end of the 15th beginning of the 16th centuries, when the population in the countryside increased, even if the economic conditions had been changing in favour of the creation of larger estates (Carnasciali and Roncaglia 1987: 10; Salvagnini 1976).

According to written sources the firsts manufacturing places for bricks and tiles were those in the environs of Impruneta, in the area of Chianti shire that we are analysing (Davidsohn 1956: 153), but they become more wide spread only from the mid 14th century onwards. During this century there is a general tendency in concentrating this kind of manufacturing activities in small and medium rural settlements: a trend mainly linked to the availability of raw materials (Mannoni 1968/69; Vannini 1977: 11-4; Francovich 1982: 20). Evidence of kilns activities in the area is widely present in documents since the central decades of the 14th century and they increase considerably during the following one. Sometimes it is not easy to locate these sites precisely so only in a few cases the dumps of kiln waste have been identified and investigated.[23] Despite the lack of archaeological assemblages already studied, both archaeological and written sources shown that tempered

pottery requiring only one firing (i.e., not glazed) could be realized in the same kilns where bricks and tiles were fired (Vanni Desideri 1982: 194-5; Francovich 1982: 22, Carnasciali and Roncaglia 1986: 24 note 6). This is not surprising as the production technique requires the same technology devices for both kinds of products.

Moreover, pottery makers were often listed in the same Guild of kiln men manufacturing building products.[24] None of them had its own Guild, but they were into the one of Stone-masons and Wood artisans[25] (Vallacchi 1991: 49); unfortunately their statues is not preserved in Florence: we only have the list of members for the year 1358 and 1512, moreover no account book of kilns and manufacturing places have survived or is knows as far as now (Carnasciali and Roncaglia 1986: 15). So far the economic and social condition of the production have been considered, but in order to elaborate a model, the availability of natural resources and the environmental conditions should be considered as well.

Written sources from the late 17th century confirm the difficulties faced in case of showers and wheat weather, described as conditions which do not allow the clay to dry properly before being fired and could cause inconvenience during the firing process as well. Despite that, winter months from December to February are sometimes indicated by written sources as a period of intense bricks manufacturing; this should be related to the fact the crafts men were usually peasants integrating their income with other activities during the months in which they did not have to work in the fields (Mazzi and Raveggi 1983: 24-8). This condition should be regarded as average (Goldthwaite 1980, Francovich 1982: 21) and it kept the same not only until the industrial revolution (Pinto 1984, Morassi 1987), but it was still the same even afterwards, during the 19-20th centuries (Quirós Castillo 2003: 394). These data about winter production might seam in contrast to what recommended by Vitruvio in the *de Architectura*: the Roman scholar indicated spring and autumn as the best time of the year for making clay artefacts as both heat and humidity could damage the objects during the drying process. He states as well that normally up two one year was needed to dry tiles nicely (Biffi 2002: 7.30-6).[26] Some scholars have suggested that

importance of kilns activity in the area around Siena made the Commune having regulations concerning it (Vallacchi 1991: 48-9).

[20] The case was the same not only in Tuscany, but in other regional areas and towns as well (Vaquero Piñeiro 2002: 144).

[21] During the 14th century several kilns in the environs of Siena belonged to religious Orders or hospitals, but there are examples of kilns being part of private estates (Balestracci and Piccinni 1977: 66). Nevertheless, it can be assumed that the manufacturing activity was mainly linked to the town estate; this could be regarded as a cause of the lacking of a strong and stable Guild (Balestracci 2000: 422).

[22] The complete study of place-names in Florence Chianti-shire is still in progress. For the Seines part it has been completed by Passeri (1983: 94, 170-3, 203, 325).

[23] It would be of great interest to investigate if, as in other context, tiles and brick wastes were sold and used for purpose like the filling of building foundations (Moorhouse 1988, 41; Moorhouse 1981: 1079). Cases of selling second choice or damaged vessels for filling vaults are well known (Vannini 2001; Francovich and Valenti 2002).

[24] At the same time stone-masons were allowed to built their own kilns to make lime and bricks (Vallacchi 1991: 49).

It would be of great interest to compare the economic and social situation of potters and kiln men devoted to tiles production, comparing as well in which Guild they were listed in different towns such as Florence, Siena and Pisa, in order to have a complete picture of all medieval Tuscany. In Siena the Guild of stone-masons was split from the one of brick and lime makers only in 1491 (ASS, *Biccherna*, 974). In Rome workers producing lime did not have their own Guild as well (Cortonesi 2002: 121), but they were listed in the "kiln men" one, which was established in 1448 (Vaquero Piñeiro 2002: 143).

[25] i.e., "Maestri di pietra e del legname".

[26] This is probably to apply to southern Europe climate condition. In England as well as in northern Europe bricks were not fired in winter because of the bad weather, but they were not even manufactured for the same reason and because of climate conditions several months might be necessary in order to dry them (Goldthwaite 1980: 188). Seasonal variations in climate could have influenced the production in Italy as

in Italy as well as in northern Europe the best time for producing and firing earthen-ware objects would have been summer (Parenti and Quirós Castillo 2000: 227-8), while others referring to Vitruvio think that spring and autumns would have been the best times (Balestracci 1984: 134-9; Carnasciali and Roncaglia 1986). Unfortunately, account books did not survive even for those 19th century farms where the manufacturing activity was still going on and, as oral tradition seams to confirm, it was linked to a rather ancient tradition. These written sources would have been quite important in order to reconstruct, by the means of bills, the variation in production during different months of the year. It would be quite interesting to make some experiments in this respect in order to have a better perception of this data.

1.2. Production models.

As discussed above, in Tuscany the production activities connected to building renewal and erection was on local-scale, but the use of bricks was wide spread even if there were sub-regional differences in the request and in the type of materials employed. These conditions can be compared to the ones existing in Liguria, a regional area which has been studied fairly in details. In Liguria the demand of bricks is more limited, but the production devices are more homogenous; this might be linked to the fact that materials travelled longer distances and by sea as well (i.e. from Genoa to Savona, **Fig. 1**), whereas in Tuscany each town controlled its production system (Parenti and Quirós Castillo 2000: 226). As discussed above, the sub-regional differences existing in Tuscany concerning the dimensions of bricks are linked to the highly fragmented political and economic conditions (Quirós Castillo 1996: 44) and it has convincingly be argued that the grade of similarity is proportional to those of economic interaction (Kula 1987) as the archaeological evidence for Genoa and Savona seem to confirm.

Archive documents have shown that workshop were spread all around the region, suggesting a local-scale production (Quaini 1972; Restagno 1980); archaeological excavations and surveys have confirmed what known from written sources (Benente and Biagini 1990; Restagno 1980; Ciciliot 1985; Albisola 1975) and proved how different production centres like Genoa and Savona influenced each other. In Liguria, as well as in Tuscany, only a few structures survived and they date mainly to the 17th-18th centuries, usually dumps of kilns waste are known from archaeological excavations (Benente and Biagini 1990: 185 note 5) rather than the kiln itself. In Liguria the structures found in Varazze (Benente and Biagini 1990, 1991) and Rossiglione (Albisola 1975) are

probably among the best preserved ones. During the 16-17th centuries rural settlements and premises were largely made out of bricks both in Tuscany (Carnasciali and Roncaglia 1986) and in Liguria (Ferrando Cabona and Mannoni 1988).

Turning to consider the area we are analysing (Chianti shire), it is worth mentioning that, as discussed above (§1.2), in Tuscany existed differences on a sub-regional scale; despite that the production models can mainly be identified as "sub-urban" with rather continuous demand, showing only a partial coincidence with "group 3" as presented in Peacock (1979) scale of producing models (Quirós Castillo 1996: 42). It is worth remembering that this scale, as any proposed model is the result of an "idealisation" (Darvill and McWhirr 1984: 247). In studying the Roman period, production characteristic and site types have been linked to the level of demand as well; according to this model a high request can be connected to cities and large towns only, valuing the demand variation as a long-term fluctuation. According to Peacock, medium demand is linked to the economy of small settlements and can be influenced by the fluctuations in local request as well as by seasonal variations. Low demand has been connected with villas and farms, with variations usually depending on building phases (Darvill and McWhirr 1984: 242). When considering this model, we should be aware that in the late Middle Ages the economic system was different than in the late Roman Empire; nevertheless, the characteristics of medium and low demand as described by the scholar, are very similar to those archive sources refer to, which have been confirmed by archaeological evidence as well. The manufacturing structures present in Florentine Chianti shire, can be referred to medium and low demand level and are not linked to towns producing system and control. Fluctuation are due both to local request and to seasonal variability; artisans could be peasants at the same time,[27] but this crafts was regarded as a regular source of income, as it was subject to tax payment (Vaquero Piñeiro 2002: 139; Goldthwaite 1980: 294).[28]

2. Manufacturing structures: characteristics and location

Concerning kilns it is worth remembering that some structure could serve several purpose and been employed not only for firing bricks, but for burning limestone and for firing pottery as well. As discusses above, the ideal place for a kiln was close to clay supplies as well as to fuel, meaning that wood was regarded as one of the most important resources.[29] In some places the cut of the wood

---

well: in central Italy a kiln could be used up to every 3 weeks and the firing lasted a few days, this resulted in 16 times firing/year. Conditions could be quite different in northern Italy, where it is known that some kilns in Mantua could have only 3 producing cycles each year (Goldthwaite 1980: 187). The production process have been studied in Rome as well and it resulted that the major part of the work was concentrated in spring and summer (Vaquero Piñeiro 2002: 146).

[27] The seasonal fluctuation of the request could result as well in a variable number of people working in the kilns. (Vallacchi 1991: 51).
[28] In this respect it is worth noting that during the 14th century the attention of the fiscal system progressively concentrated on craftsmen (Vaquero Piñeiro 2002: 142).
[29] The use of wood and the reduction of forest is constant during the Middle Ages and linked to the manufacturing activities of the 14th

was planned by the whole community and was regulated by rigid rules, sometime resembling a ritual.[30] The presence of the wood combined with clay supply is often mentioned as a great advantage in archive sources describing properties (Benente and Biagini 1990: 189), but reducing transport prices was important when trading the final products as well. Because of that small kilns were built in rural context only when needed, but the production could die out and be started in another sites once it was not longer necessary (Restagno 1980). This was not the case when the manufacturing process was done in a large estate or was under the control of towns (Balestracci 2000).

As a consequence, the kiln was the closest as possible both to raw material supplies and to the building site as well. It was quite usual that small kilns for making bricks and lime were built on the yard: examples are known from written sources in Siena as well as in Florence (Mazzi and Raveggi 1983: 267 note 55); some structures could produce only one kind of artefacts, but in some cases different kind of bricks could be made and the production was not a short-term one (Corsi 1991: 25).

In Florentine Chianti shire only a few archaeological excavations have been undertaken; but surveys have shown the presence of different type of structures. Kilns or dumps and archaeological assemblages which can be related to bricks, lime or pottery production investigated as far as now are presented below in order to give some key-examples of different manufacturing process involving pyrotechnology. Archive sources show that some kilns produced not only pottery, but bricks and roof tiles at the same time as well; moreover there were families producing earthenware manufactures till the 19th century (Impruneta 1980: 163).

2.1 Brick kilns

Archaeological evidence show the wide distribution of bricks production (Goldthwaite1980: 172) and recently further structure have been identified. Concerning brickworks it can be said that the manufacturing structure included not only the kiln itself, but clay pits, spaces to model and mould clay and premises to stock the finished products. The process of clay preparing requested an open space and some premises, like a porticos (Goldthwaite1980: 181-2)[31] where cleaning it from stones and other impurities, a place were to dry the bricks before

firing them,[32] (a process which usually requested several weeks time) and a stocking area where to keep finished products as well (Carnasciali and Roncaglia 1986: 20). The perimeter of the kiln was usually rectangular: two,[33] but sometimes three "galleries" where made by building vaults resulting from arcades elevated at a regular distance. The space below the arcades was then used as a burning chamber and the one above for charging the bricks to be fired. Similar structures are known not only in Tuscany, but in Liguria[34] and in Umbria as well.[35] Sometimes the two parts could be separated by a floor made with pierced tiles, but this device, even though wide spread in Roman times and still in use during the early Middle Ages (Moran 2000: 171) was hardly ever employed in the late Middle Ages and in post-Medieval times, when only a few examples are known as far as now (Benente and Biagini 1991). In some cases the bricks employed for building the kilns were reused and resulted from the pillages of ancient structures, but the study of bricks dimensions and shape combined with chemical analyses on clay, have proved that in some cases they had been manufactured by the same artisans and using the same clay supplies (Benente and Biagini 1991: 186-7). If the latter is the case, the bricks made for constructing the kilns might have been burned using a simpler structure, such as a clamp-kiln. Despite the latter could be regarded as "rudimental" and was used in some regional areas in Italy during the Middle Ages only[36] and for limited and occasional productions, was still employed in Germany and in England up to the 19th century (Corsi 1991: 23).[37]

---

[32] In 1540 Biringuccio dedicated a whole chapter of his book to pottery making (Book 9) and another one to bricks and lime manufacturing (Book 2). When talking about vessels making, he describe the process of preparing the clay, which requested to take away any small stone or impurity and to dry the objects once done (Biringuccio 1540: 145).

[33] Two vaults were present in early-medieval brickworks as well, and resulted from three parallel walls wide 55 cm and long 5 m, covering a surface of about 20 square metres (Moran 2000: 171).

[34] This is the case for a brick kiln excavated in Liguria and known from written sources too (Albisola, 1975: 151-2).

[35] There are several examples of this kind of kiln in the region, with a firing chamber made of arches built at a distance of 20-25 cm, 1 m high and 1.1-2 m wide. Some of them dating from the 14th century were still in use during the 18th (Covino and Giansanti 2002: 14, 62, 94).

[36] In Rome this way of firing bricks or producing quicklime was not used as there was not enough wood available and the final products were of lower quality, moreover, fearing the damages which might have occurred in case of an excessive wood exploitation the production of lime was regulated and fix amount were established (Cortonesi 2002: 129, 113).

[37] In late 19th century England a clamp was still regarded as more convenient than a kiln. This could be explained with the large amount of wood and coal available, while in Italy fuel was very expensive and brushwood and light timber not available for the whole year. Permanent kilns allowed to save fuel and required less time (Goldthwaite 1980: 186). "In medieval Europe bricks were commonly burnt in a clamp (or heap), a largely temporary structure that was built to be fired and dismantled after each firing. With ashes mixed into the clay so that the bricks created their own heat and with fuel packed between that stacks of bricks, a clamp simply burned itself out. Although kilns were not unknown, clamps were used extensively throughout the early modern period, the technique being the normal procedure for making bricks even in England and Germany well into the nineteenth century." (Goldthwaite1980: 177).

---

century (Lusini 1904), but it would become on wide scale only during the 16th (Balestracci 1984: 133; Cherubini 1974: 43; 1981: 304, 309); when possible, wood was used rather than coal (Rauty 1987: 143). It has been suggested (Balestracci 1984: 133) that the model proposed by Devèze (1961) for the exploitation of wood resources in France could be applied to Europe as well (Cipolla 1974: 285; Sereni 1974: 132; Braudel 1982: 333).

[30] In Strada in Chianti the wood was cut every 5 years (Carnasciali and Roncaglia 1986: 22).

[31] In Archive sources dating to the 18th century (Cabrei) kilns are usually represented with a porticos, which was used to dry the artefacts (Albisola 1975: 154).

It is worth mentioning that while in the early Middle Ages reusing Roman bricks was quite usual, this became less frequent after the 11th century, when the manufacturing system started to be vital again. Nevertheless it is rather easy to distinguish a Roman brick from a Medieval one, whose devices might have originated in northern Italy during the 9th century (Mannoni 1984: 396),[38] but the area from where the technical skills spread and made possible to restart a large-scale production after the 11th century, is still to be identified (Quirós Castillo 2003: 389).[39] Talking from a production point of view, it seams like there are some differences in bricks and roof tiles production: in fact the latter seams to have been produced continuously (Parenti and Quirós Castillo 2000: 220-1).

The kiln investigated in Lamole (**Fig. 2**) is not isolated, but built next to a farm-house and on a hillside and has premises like a small porticos on the front (**Figs 3-4**) – usually mentioned in documents – on the front and some storing rooms next to it. The hill is built on has natural sediments of clay and there are two wheel next to it[40]. The firing chamber is vaulted and results from bricks arches at a regular distance (**Figs 5-6**); there is no floor separating it from the upper chamber and it has vertical walls forming a flue which is covered by a roof hold by small brick pillars (**Fig. 3**); next to the upper part there is another portico. Building kilns on hillsides made easier to load them from the higher up ground level (Goldthwaite 1980: 187). The presence of pillars holding the roof of the cooking chamber are known in Rome and its surroundings in structures dating from the mid 16th century onwards (Giustini 1997: 10, 14); they are similar to those described in a treaty of the early 19th century (Valadier 1828: 42) and some of them were still in use at that time, showing the long-term tradition of the craft (Giustini 1997: 44, 53). The basement of the chamber was usually cut into a natural clay embankment. In some cases, like the kiln found in Varazze (Liguria), only one part had two different chambers, while firing by direct contact with the flame was limited to only one third of it (Benente and Biagini 1991; Piccolpasso and Conti 1976: 112)

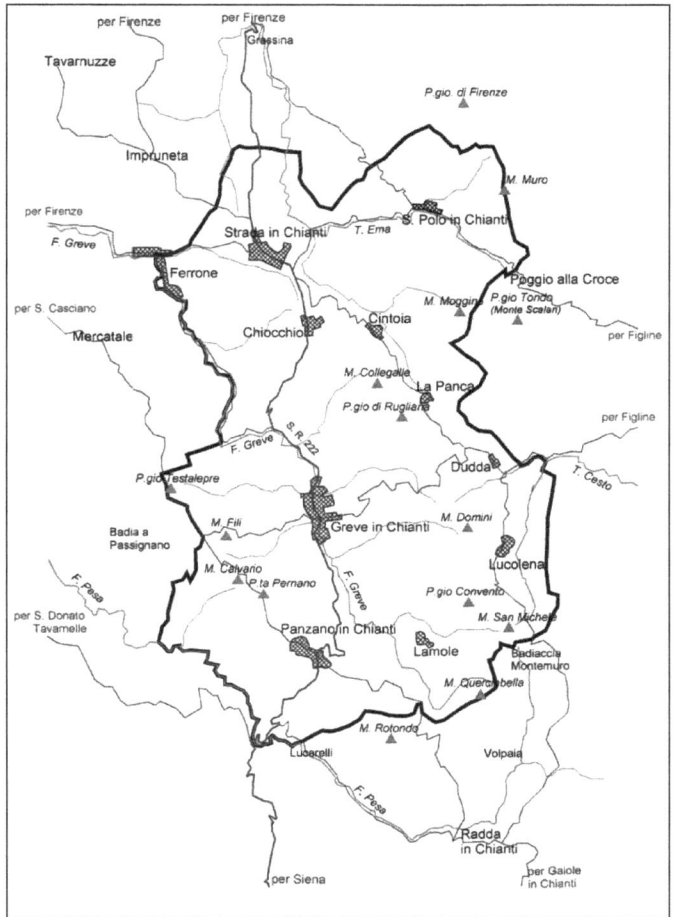

Fig. 2: Greve in Chianti and its environs: kilns sites.

Fig. 3: Bricks kiln in Lamole: premises

---

[38] Dimensions are usually 30-35 x 14-16 x 6.5-7 cm (Corsi 1991).

[39] For bricks and tiles there is not continuity in production from Roman times to the Middle Ages, the only exception are the area under the Byzantine control. The first Tuscan buildings made of bricks with Medieval characteristic date from the 12th century onwards; in this case as well there are exceptions: the dome of the Cathedral in Pisa (1064-1100) is made with Medieval bricks, but their dimensions have not been studied yet (Quirós Castillo 2003: 390).

[40] Kilns were usually located on hillsides, with the fire chambers at the lower level in front, partly to help increase the draught, and with the ovens above therefore accessible for loading from the ground level higher up the slope in back.

Fig. 4: Bricks kiln in Lamole: porticos and entrance.

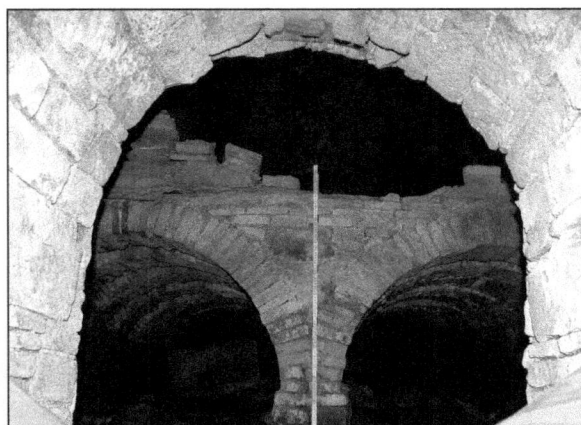

Fig. 5: Bricks kiln in Lamole: burning chamber.

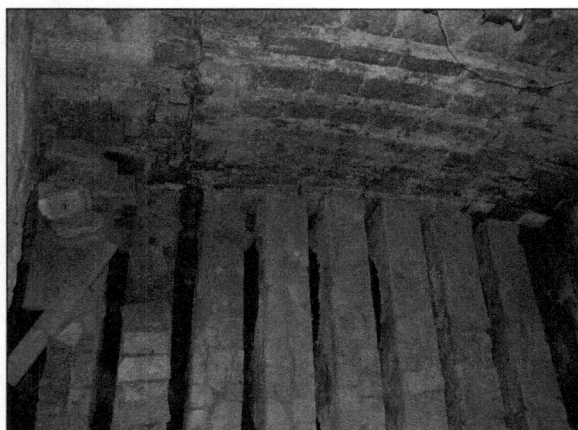

Fig. 6: Bricks kiln in Lamole: the interior.

Pottery kilns had a similar structure, but the perimeter could be circular or squared instead of rectangular and they were covered with a vault, whereas brickworks were open on the top and were covered with tiles dumps, old roof tiles and sometimes slabs. (Benente and Biagini 1991: 188). They were similar in the respect that the chamber was dug in the ground (Thiriot 1995). The structure of the bricks' kiln found in Lamole, resemble the one described by Piccolpasso and Conti (1976: 122-4) for pottery making in the mid 16th century,[41] but lack of the floor. It resemble as well the one excavated in a small village in Liguria (Rossiglione), which is mentioned in a notary's document dating 1744 (Goldthwaite1980: 182) and is a kind of structure which was quite spread in northern Italy from the 16th century onwards (Catarsi dall'Aglio 1996: 760), but there are earlier examples dating before the 14th century (Baldassarri *et al.* 2005: 88). This kind of structures was still in use between the 17th and the 19th century (Restagno 1976). The research on archive sources referring to this area is still in progress, but oral sources confirm the presence of a manufacturing structure at least since the beginning of the 18th century. The similarities between the two kilns are so strict that they might have been built in the same period (Albisola 1975). Other examples are known in Umbria as well, where a research on this topic involving oral sources, archive documents and archaeological evidence has been recently carried out. The presence of this kind of buildings is known since the 13th century, when they are often mentioned in the description of maps. Oral sources, as well as archaeological evidence[42] have confirmed that some of them were still in use in the late 18th century, showing a long-term tradition both in manufacturing activities and in exploiting natural resources (Covino and Giansanti 2002: 62, 94).

2.2 Lime kilns

As for brick-making limestone could be converted into quicklime without using a kilns structure, but by preparing a pit and filling it with a hip of material and

---

[41] *Le furnace adunque, la più parte, dico io, si fanno di matoni crudi, a guisa di camerette; vero è che una parte ne viene sotto terra, e questa è quella dove stanno le bragie. Dico che vien cavata sotto da un piede o un piede e mezzo. Se ne fanno de grande e de piccole. [...]Quelle che usam noi si fanno 5 piedi larghe e 6 alte et altritanto lunghe, e 4 piedi alte sotto gli archetti. [...] Molti gli mattoni, che vano da l'un arco e l'altro, cavano da tutt'e dua le bande [...] quai raggiunti insiemi, lassano di aperto un foro perfetto, come qui, e questo si fa per gli saglimenti del fuoco. Altri sogliano far questi salimenti con lassare gli mattoni alquanto uno discosto da l'latro, e questo è più in usso.* Piccolpasso and Conti (1976: 122-5). The major part of the kiln is made with unfired bricks and it looks like small chambers one next to the other. A part is below the ground and is in there where the fire is set. The chamber is usually dug in the ground for one feet or for one and half feet. There are big and small kilns. The ones we use are 5 feet wide and 6 feet high and they measure the same in length; they are 4 feet high below the arches. The brick joining the two sides of the kiln are pierced, so that joined together they leave a round hole through which the fire goes up. Other people make this holes by leaving some space between two bricks and this is the most common way of doing it.

[42] One of these kilns shows sign of restoration and in the upper part of the flue there is a brick dated 1775 (Covino and Giansanti 2002: 94).

constructing around it a clamp of wood covered with clay and heart and then setting it to fire. One of the first description of this rather simple kiln in Renaissance texts is the one by Biringuccio in 1540[43], but Roman texts give similar description as well (Catone, *De Agricoltura XLIV*). As it will be discussed below when presenting the archaeological evidence from the Castellaccio di Lucolena this manufacturing process was still in use during the 19th century and could be employed to fire either lime or bricks, but needed to be dismantled after each firing process (Goldthwaite 1980: 186-7).[44] The process was the same as the one for making coal out of wood. When a structure was used, this was made of a sole chamber containing both limestone to be burned and fuel (usually wood). Stones rich in calcium carbonate were tipped from a hole on the top and there was a stock-hole as well which allowed to put new fuel in if necessary. As the structure was a simpler one than those requested for firing bricks and premises were not needed, lime kilns were regarded as less valuable.[45] Nevertheless, owning a limekiln was regarded as a sign of relatively richness, as only well to do people sometime had manufacturing structures such as a kiln for making lime in their properties in the countryside (Mazzi and Raveggi 1983: 148). As for craft linked to clay exploitation, the making of quicklime was a way of increasing the income for the rural community and even if not produced in the same amount during the year, it cannot be regarded as an occasional activity (Balestracci 1984: 136-7). Statement of tax paying confirm for lime, as well as for bricks and pottery, that the manufacturing process was a source of income (Vaquero Piñeiro 2002: 140).

There are plenty of examples of lime kilns either built in central towns like in Rome, or constructed along the main road or next to monumental sites like in Ostia (Lenzi 1998), so to exploit the archaeological remains. This was the case for a small village called Lucolena, located in the environs of Greve in Chianti (**Fig. 2**), where a vaulted kiln was built next to the main road and only one mile away both from the ruins of a medieval castle and from the raw material supply. The chamber is dug in the ground and even though dating several centuries later

Fig. 7: Lime kiln in Lucolena: the interior.

(probably 17th), the whole structure show similarities with the one constructed in 11th century Pistoia, where lime was manufactured on a local scale, but it was enough to supply the building of the Bishops' palace (Rauty 1987: 139-140). In Pistoia the kiln was built using older structure as walls and digging a pit as a firing chamber, the raw material was then pilled from the top; the upper part is not preserved, but it is likely that it was vaulted (Vannini 1977: 187-9). Despite the similarities the kiln in Lucolena (**Fig. 7**) was built next to natural resources; as it will be discussed below, this meant reducing carriage costs: in fact, transporting quicklime slabs was easier than moving big blocks of limestone. It is worth noting that these type of kilns show several similarities with those present in northern Europe and a continuing tradition from Roman times onwards (Baragli 1998: 129); in some cases there are similarities with those for lime and bricks making dating from the early Middle Ages (Otranto 1992; Castellana 1994).

2.3 *Lime and bricks manufacturing in clamps*

As already discussed above, structures for manufacturing bricks and lime could be the same, having sometime the

---

[43] Biriguccio states that lime is made in a cave where a round pit is dug. The pit should be as big as many stones one wants to allow in, "*et primamente per fare quella della calcina si fal in una grotta una fossa tonda cavando all'ingiù, di forma quasi ovale, qual sia di tanta capacità che il vacuo contenga la quantità che n' volete*" (Biringuccio 1540: 147 v).

[44] While in Florence producers of bricks and lime were often the same ones (Goldthwaite 1980: 267-8), in Rome they were specialised in one manufacture (Vaquero Piñeiro 2002: 153). The major difference was the one existing between those who were employed in the manufacture and those who controlled the structure and the production activity (Vaquero Piñeiro 1996: 479); moreover during the 17th century, as the building activity was still one of the most vital ones in Rome, potters were forbidden to produce bricks and lime and some regulations in this respect were introduced (Vaquero Piñeiro 1996: 485 note 86)

[45] Referring to a lime kiln in Impruneta and dating to the mid 15th century Goldthwaite states (1980: 181). "Kilns where only lime was burned were much less value than brick kilns; in the tax records for Impruneta, an important center for the production of floor and roof tiles, the assessed value of most kilns was a modest 30 to 40 florins."

latter a more articulated structure (Vallacchi 1991: 41). In Lucolena the kiln for lime-making described above was not the only one, but lime slabs once used for roofing the Medieval Castle (11-14th centuries) were burned and turned into quicklime on the archaeological site as well by using a clamp. As the clamp kiln needed to be destruct, the dumps only are preserved (**Fig. 8**), and show that both limestone and bricks where fired on the site; as there was plenty of wood available in the surroundings and the final products was lighter than the stones and easier to be transported, this resulting in a reduction of carriage costs. The assemblage has only in part been excavated and the research is still in progress and in some respect at a starting point. As already stated above, in Pistoia lime-stones were carried on the building ground and cooked in there, instead of transporting quicklime ready to be used and cooked next to the raw material supplies as it usually happened (Baragli 1998: 126). Further research on market, viability and on the dynamic of raw material supply in the environs of Pistoia are needed in order to understand these differences. Nevertheless, the two examples show some similarities: contracts for rending some premises in order to "dig" old stones are known in Pistoia (Rauty 1987: 141-2) and for the Castellaccio oral sources refers the same conditions, in fact lime manufacture exploiting the ruins of the castle was regulated by an agreement between the stone-masons working in there during the 18-19th centuries and the owner of the ground.

Fig. 8: Castellaccio di Lucolena: dumps of clamp kiln (transect).

2.4 Pottery

Concerning pottery there is a long-term tradition linked to the technique of moulded manufacture, used both for making oil jars and vessels for storing cereals and for small high-decorated bowls as well. In some cases the production lasted only a few decades before dying out, but in others kept on till nowadays. Referring to the first one it is worth mentioning the late 14th -beginning of the 15th century, moulded pottery known as "Figlinese" (**Fig.**

9) ,[46] already studied in the past and recently taken again into account considering the latest archaeological evidence.[47] Regarding the latter oil moulded jars and big lemon vessels are still manufactured in Ferrone and Impruneta. Pottery manufacturing will not be discussed in details in this paper, what is important to point out is that producing tempered vessels such as big oil jars requested the same technical skills involving pyrothecnology as for firing bricks.

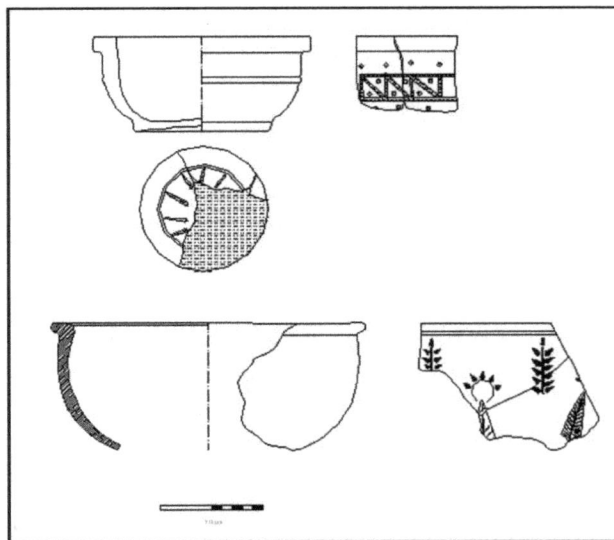

Fig. 9: "Figlinese": moulded bowls.

3. Discussion

When trying to reconstruct a producing model, it is important to consider that both technical devices and economic models should be taken into account; moreover, different models were depending on demand and they could change from place to place relating to conditions. The presence of natural resource was an essential element in deciding the setting of a craft activity. Clay and wood supplies, which represent the raw material and the fuel employed should be in the environs of the working premises. As carriage costs were one of the most expensive things in this economic system, lime and brick kilns were located in a place where it was not necessary to transport the final products far away.[48] When

---

[46] The name of this type of pottery derive from the village of Figline di Prato, where a kiln was investigated for the first time (Maetzke 1973). The meaning of the place-name "Figline" as deriving from the latin figulus (pottery makers) referring to a site were vessels were manufactured has been discussed by Carnasciali and Roncaglia (1986: 24 note 4), who refers to both Cora (1973) and Ballardini (1964: 247)

[47] One paper has been presented recently at the annual meeting of the MPRG in Chester (2006): Storage, cooking and display pottery from two fortified settlements in Chianti: Castellaccio di Lucolena (10th - 13th cent) and Monte Moggino (14th-15th cent).

[48] It is worth mentioning that the first organised brickworks which appeared after the end of the Roman Empire were mainly connected to the building and expansion of important monasteries, which started

possible, quicklime rather than limestone were transported on the building site, but the kilns was in the immediate nearby in both cases. The same structure could sometime been used both for manufacturing lime and bricks, but clamps were still used up to 19th century as brooms and wood were widely available in great amount.

Acknowledgements

I am indebted to Andrea Garuglieri, friend and president of the Archaeological Section of the volunteers association "GEV" (Greve in Chianti) who has been carried out an extensive research for locating archaeological sites in Florentine Chianti shire during the last decade and drawn Figure 2. My gratitude goes as well to the friends and members of the association, especially to Andrea Gigliotti who took part in collecting the first data on the kiln in Lamole and made some preliminary drawings. I am very grateful to Paolo Socci (Fattoria di Lamole – Greve in Chianti) and to the Tramonti family (Fattoria dell'Ottavo – Lucolena) for the access to their properties and for the interest and care shown in preserving the archaeological evidence. Not only my gratitude but those of all members of the GEV goes as well to Carlotta Cianferoni and Anna Rastrelli of the National Heritage (Soprintendenza Archeologica della Toscana – Florence) for having encouraged the survey research and the emergencies excavations carried out in the last few years.

Bibliography

ALBISOLA
  1975 *Rinvenimento di una fornace di mattoni a Rossiglione Inferiore*, in *Atti dell'VIII Convegno Internazionale della Ceramica*, Albisola: Centro Ligure per la Storia della Ceramica, pp. 151-4.
ASS (Archivio di Stato di Siena) *Biccherna*, 974 (Siena National Archive).

BALDASSARRI, M., M. FEBBRARO, M. MESCHINI, A. MEO, S. SACCO, and I. TROMBETTA
  2005 La produzione di laterizi e l'edilizia in cotto nel Valdarno Inferiore medievale: il caso di Marti (Pisa), *Archeologia Medievale* 32: 77-96.
BALESTRACCI, D.
  1984 La zappa e la retorica. Memorie familiari di un contadino toscano del Quattrocento, *Quaderni di Storia Urbana e Rurale* 4: 134-139. Firenze: Salimbeni.
  2000 Produzione ed uso del mattone a Siena nel medioevo, in P. Boucheron (ed.), *La brique antique et médiéval. Collection de l'école Française de Rome* 272: 417-428.
BALESTRACCI, D. and G. PICCINNI
  1977 *Siena nel Trecento. Assetto urbano e strutture edilizie*, Firenze: Edizioni CLUSF.
BALLARDINI, G.
  1964 *L'eredità ceramistica dell'antico mondo romano*, Roma: Istituto Poligrafico dello Stato.
BARAGLI, S.
  1998 L'uso della calce nei cantieri medievali (Italia centro-settentrionale): qualche considerazione sulla tipologia delle fonti, *Archeologia dell'Architettura* 3: 125-139.
BENENTE, F. and M. BIAGINI
  1990 Scavo di recupero di una fornace per laterizi a Varazze (GE), *Archeologia Medievale 17*: 347-354.
  1991 Una fornace per laterizi del XVII secolo a Varazze. In *Atti del XXIV Convegno Internazionale della Ceramica (1991)*, Albisola: Centro Ligure per la Storia della Ceramica. pp. 185-192.
BIFFI, M. (ed.)
  2002 *La traduzione del de Architettura di Vitruvio (ms. II.I.141 BNCF), di Francesco di Giorgio Martini*, Strumenti e Testi 9, Pisa: Scuola Normale Superiore.
BIRINGUCCIO, V.
  1540 *De la Pirotechinia*, A. Cargo (ed.) (1977), Milano: Edizioni il Polifilo.
BOTTARI SCARFANTONI, N.
  2004 La produzione di mattoni nel XIV secolo a Pistoia e il loro utilizzo. In E. Daniele (ed.), *Le dimore di Pistoia e della Valdinievole. L'arte dell'abitare tra ville e residenze urbane*. Firenze: Alinea, pp. 225-229.
BRAUDEL, F.
  1982 *Civiltà materiale, economia e capitalismo (secoli XV-XVIII)*, I, *Le strutture del quotidiano*, Torino: Einaudi.
BUSTI, G. and F. COCCHI
  1996 *Terrecotte e laterizi*. Perugia: Electa.
CALDELLI, A.
  1991 Il ruolo dei fornaciai a Siena fra Medioevo ed età Moderna. In *Fornaci e mattoni a Siena dal XIII secolo all'azienda Cialfi*, CRAM: Siena, pp. 31-40.

attracting specialised artisans. Only seldom manufacturing structure were connected to towns estate (Negro Ponzi, 2000: 54). Kilns were constructed on the building site, avoiding the costs of transports; their activity might have quickly died out once the construction phase was over, but it is possible as well that the production have been carefully planned (Moran 2000: 176-7).

While in the 5th century bricks manufacturing was a prerogative of the urban elite, since the mid 8th century the manufacturing activity started to be strictly linked to monasteries and was carried out far away from the towns. At that time the introduction of new techniques made possible the production an a wide scale such as the one in Roman times. The kilns had the same form as those used in Renaissance times: rectangular with a vaulted fire-chamber. One of the earliest examples of this sort known in Italy as far as now for the early Middle Ages seams to be the one found in San Vincenzo al Volturno, which has been explained as the result of the continues contact with the Byzantine area during the second half of the 8th century (Moran 2000: 180-1).

CARNASCIALI, M. and G. RONCAGLIA
    1986   *Antiche fornaci nel Chianti.* Radda in Chianti.
CASTELLANA
    1994   Una fornace di età bizantina a Castellana di Pianella (PE), *Archeologia Medievale* 21: 269-86.
CATARSI DALL'AGLIO, M.
    1996   Fornaci d'epoca post-antica nel territorio di Collecchio (prov. Parma), *Archeologia Medievale* 33: 755-62.
CHERUBINI, G.
    1974   Vita trecentesca nelle novelle di Giovanni Sercambi. In *Signori, contadini, borghesi. Ricerche sulla società italiana del basso medioevo*, Firenze, pp. 3-49.
    1981   Le campagne italiane dall'XI al XV secolo. In *Storia d'Italia*, IV, *Comuni e Signorie: istituzioni, società e lotte per l'egemonia*, pp. 265-448.
CICILIOT, F.
    1985   Note sulle fornaci di mattoni nel savonese. In *Atti del XVIII Convegno Internazionale della Ceramica*, Albisola: Centro Ligure per la Storia della Ceramica (1988), pp. 173-4
CIPOLLA, C. M.
    1974   *Storia economica dell'Europa pre-industriale.* Bologna: Il Mulino.
CORA, G.
    1973   *Storia della maiolica di Firenze e del Contado. Secoli XIV e XV.* Firenze: Sansoni.
CORSI, R.
    1991   Forma, dimensioni e caratteristiche del mattone senese. In *Fornaci e mattoni a Siena dal XIII secolo all'azienda Cialfi.* Siena: CRAM. pp. 21-30.
CORTONESI, A.
    1987   E sorsero le fornaci attorno alla città del Duecento, *Storia e dossier* II: 42-5. 8 giugno 1987.
    2002   Fornaci e calcare a Roma e nel Lazio: secoli XIII-XV. In Lanconelli, A.and I. Ait (eds.) *Maestranze e cantieri edili a Roma e nel Lazio. Lavoro, tecniche e materiali nei secoli XIII-XIV*: 109-136.
COVINO, R. and M. GIANSANTI
    2002   *Fornaci in Umbria. Un itinerario di archeologia industriale*, Electa: Perugina.
DARVILL, T. and A. McWHIRR
    1984   Brick and tile production in Roman Britain: models of economic organization, *World Archaeology* 15 (3): 239-261.
DAVIDSOHN, R.
    1956   *Storia di Firenze*, vol. I, Firenze: Sansoni.
DEVEZE, M.
    1961   *La vie de la foret française au 16me siècle.* Paris: S.E.V.P.E.N.
FERRANDO CABONA, I. and T. MANNONI
    1988   *Liguria ritratto di una regione. Gli edifici tra storia ed archeologia*, Genova: Sagep. pp. 15-280.

FRANCOVICH, R.
    1982   *La ceramica medievale a Siena e nella Toscana meridionale (secc. XIV-XV). Materiali per una tipologia.* Firenze: All'Insegna del Giglio.
FRANCOVICH, R., and M. VALENTI
    2002   *C'era una volta: la ceramica medievale nel convento del Carmine (Santa Maria della Scala, 25 giugno-15 settembre 2002)*, Firenze: Polistampa.
GIUSTINI, L.
    1997   *Fornaci e laterizi a Roma: dal XV al XIX secolo.* Roma: Edizioni Kappa.
GOLDTHWAITE, R. A.
    1980   *The building of Renaissance Florence. An Economic and Social History,* Baltimore and London: Johns Hopkins University Press.
IMPRUNETA
    1980   *La civiltà del cotto. Arte della terracotta nell'area fiorentina dal XV al XX secolo,* Impruneta, maggio-ottobre 1980.
KULA, W.
    1987   *Le misure e gli uomini dall'Antichità a oggi.* Bari: Laterza.
LENZI, P.
    1998   "Sita in loco qui vocatur calcaria": attività di spoliazione e forni da calce a Ostia. *Archeologia Medievale* 25: 247-263.
LISINI, A. (ed.)
    1903   *Il costituto del comune di Siena volgarizzato nel MCCCIX-MCCC.* Siena.
LUSINI, V.
    1904   *Dell'arte del legname innanzi al suo statuto del 1426,* "Bullettino senese di storia patria" IX: 183-246.
MAETZKE, G.
    1973   Una fabbrica di ceramica d'uso acroma decorata a rilievo a Figliene di Prato. In F. Guerrieri and G. Maetzke, *La pieve di Figline di Prato: Il suo patrimoni artistico,* Prato: Libreria del Palazzo: 99-114
MANNONI, T.
    1968/1969   La ceramica medievale a Genova e nella Liguria. *Studi Genuensi* 7 (1968/69), Bordighera-Genova (1975).
    1984   *Metodi di datazione dell'edilizia storica,* "Archeologia Medievale" XI: 396-403.
MANNONI, T. and M. MILANESE
    1988   Mensiocronologia. In Francovich, R. and R. Parenti (eds.), *Archeologia e restauro dei monumenti. I ciclo di lezioni sulla ricerca applicata in archeologia (Certosa di Pontignano-Siena: 28 settembre-10 ottobre 1987),* Firenze: All'Insegna del Giglio. pp. 383-402.
MAZZI, M. S. and S. RAVEGGI
    1983   *Gli uomini e le cose nelle campagne fiorentine del Quattrocento,* Firenze: Olschki.
MOORHOUSE, S.
    1981   The rural Medieval landscape. In Faull, M. L. and S. Moorhouse (eds.), *West Yorkshire: an*

Archaeological survey to AD 1500, 4 vols., Wakefield: West Yorkshire MCC. pp. 581-850.

1988 Documentary evidence for Medieval Ceramic roofing materials and its archaeological implications: some thoughts. *Medieval Ceramics* 12: 33-56.

MORAN, M.

2000 Produzione di laterizi in un monastereo meridionale in epoca carolingia: San Vincenzo al Volturno. In Gelichi, S. and P. Novara (eds.), *I laterizi nell'alto Medioevo italiano*. Ravenna: Società di Studi Ravennati. pp. 169-184.

MORASSI, L.

1987 La fornace nell'economia agricola del Settecento. In Buora, M., and T. Ribrezzi (eds.) *Fornaci e fornaciai in Friuli*. Udine: Ribezzi. pp. 95-111.

NEGRO PONZI, M. M.

2000 La produzione e l'uso dei laterizi nei siti rurali d'Italia settentrionale tra tardo antico e medioevo: i laterizi di Trino (Vc). In Gelichi, S. and P. Novara (eds.), *I laterizi nell'alto Medioevo italiano*. Ravenna: Società di Studi Ravennati. pp. 53-74.

OTRANTO

1992 Fornaci altomedievali ad Otranto. Nota Preliminare, *Archeologia Medievale* 19: 91-122.

PARENTI, R. and J. A. QUIRÓS CASTILLO

2000 La produzione dei mattoni della Toscana medievale (XII - XVI secolo): un tentativo di sintesi. In Boucheron, P. (ed.), *La brique antique et médiévale*. Roma: École Française de Rome. pp. 219-235.

PASSERI, V.

1983 *Repertorio dei toponimi della provincia di Siena desunti dalla cartografia dell'Istituto Geografico Militare*, Siena: Amministrazione Provinciale di Siena.

PEACOCK, D.

1979 An ethno archaeological approach to the study of Roman Bricks and Tiles. In McWhirr, A. (ed.) *Roman Brick and Tile. Studies in Manufacture, distribution and use in the Western Empire*, BAR International Series 68: 5-10.

PICCOLPASSO, C. and G. CONTI (eds.)

1976 *Li tre libri dell'arte del vasaio*. Firenze: All'Insegna del Giglio.

PINTO, G.

1984 L'organizzazione del lavoro nei cantieri edili (Italia centro-settentrionale). In *Artigiani e salariati. Il mondo del lavoro nell'Italia dei secoli XII-XV. Atti del decimo convegno internazionale (Pistoia, 9-13 ottobre 1981)*, Pistoia, pp. 69-101.

QUAINI, M.

1972 La localizzazione delle fornaci savonesi in una prospettiva geo-storica. In *Atti del IV Convegno Internazionale della Ceramica*, Albisola: Centro Ligure per la Storia della Ceramica, pp. 299-310.

QUIRÓS CASTILLO, J. A.

1996 Produzione di laterizi nella provincia di Pistoia e nella Toscana medievale e postmedievale, *Archeologia dell'Architettura* I: 41-51.

2003 Mattoni nella Toscana medievale: dimensioni e cronologia. In Badstübner, E. (ed.) *Backsteintechnologien in Mittelalter und Neuzeit*. Berlin: Lukas-Verlag. pp. 388-402.

RAUTY, N.

1987 Tecniche di costruzione e di cantiere nell'antico Palazzo dei Vescovi di Pastoia (secoli XI-XIV). In *Undicesimo Convegno Internazionale. Tecnica e Società nell'Italia dei Secoli XII-XVI*, Pistoia: Centro Italiano di Studi di Storia e d'Arte, pp. 135-156.

REPETTI, E.

1833-1845 *Dizionario geografico, fisico, storico della Toscana*, I-VI, Firenze: Tofani-Mazzoni.

RESTAGNO, D.

1976 Primi risultati di un'indagine sulle antiche fornaci di Albisola. Le fornaci di Albisola Superiore e della frazione Capo tra XVII e XIX secolo. In *Atti del IX Convegno Internazionale della Ceramica*, Albisola: Centro Ligure per la Storia della Ceramica. pp. 351-384.

1980 Fornaci di mattoni nella valle del Riabasco. In *Atti del XIII Convegno Internazionale della Ceramica*, Albisola: Centro Ligure per la Storia della Ceramica, pp. 279-286.

RICE, P. and W. KINGERY (eds.)

1997 The Prehistory and History of ceramic kilns, *The American Ceramic Society*, 7.

SALVAGNINI, G.

1976 Agricoltura e case rurali in Toscana alla fine del Cinquecento, *Granducato*, 6, Firenze: Osservatorio fiorentino di storia, arte, cultura.

SERAFINI, L.

2003 La costruzione in laterizio: materiali, forme, tecnologie in Abruzzo. In Fiengo, G. (ed.), *Atlante delle tecniche costruttive tradizionali*. Napoli: Arte Tipografia Editrice. pp. 165-174.

SERENI, E.

1974 *Storia del paesaggio agrario italiano*. Bari: Laterza.

THIRIOT, J.

1995 Les ateliers. In *Le vert et le brun de Kairouan a Avignon, céramique du Xe au XVe siécle*, Musées de Marseille – Réunion des Musée Nationaux, pp. 19-40.

VALADIER, G.

1828 *L'architettura Pratica*, vol. I, Rome.

VALLACCHI, F.

1991 Dislocazioni della fornaci e loro continuità d'uso sul territorio dello stato senese tra medioevo ed età moderna. In *Fornaci e mattoni a Siena dal XIII secolo all'azienda Cialfi*, Siena: CRAM, pp. 41-52.

VANNI DESIDERI, A.
1982 Fornaci e vasellai in un centro minore del basso Valdarno, *Archeologia Medievale* 9: 193-216.

VANNINI, G.
1977 *La maiolica di Montelupo: scavo di uno scarico di fornace.* Montelupo Fiorentino: Rinascita.
2001 Una struttura edile trecentesca: il complesso fittile di San Domenico in Prato. In De Minicis, E. (a cura di), *I laterizi in età medievale. Dalla produzione al cantiere.* Atti del Convegno Internazionale di Studi, Roma 4-5 giugno 1998, pp. 199-212.

VAQUERO PIÑEIRO, M.
1996 L'Università dei Fornaciai e la produzione di laterizi a Roma tra la fine del '500 e la metà del '700, *Roma moderna e contemporanea* 4: 471-494.
2002 La gabella dei calcarari: note sulla produzione di calce e laterizi a Roma nel Quattrocento. In Lanconelli, A. and I. Ait (eds.) *Maestranze e cantieri edili a Roma e nel Lazio. Lavoro, tecniche e materiali nei secoli XIII-XIV.* Roma: Vecchierelli. pp. 137-154.

WICKHAM, C.
1988 L'edilizia dell'alto medioevo, *Archeologia Medievale* 15: 105-124.

ZDEKAUER, L.
1897 *Biccherna* in *Il costituto del Comune di Siena dell'anno 1262,* Milano.

www.ingramcontent.com/pod-product-compliance
Lightning Source LLC
Chambersburg PA
CBHW061002030426
42334CB00033B/3330